CHINA'S STRATEG
IMPLICATIONS FOI

Mark A. Stokes

September 1999

*****

The views expressed in this report are those of the author and do not necessarily reflect the official policy or position of the Department of the Army, the Department of the Air Force, the Department of Defense, or the U.S. Government. This report is cleared for public release; distribution is unlimited.

*****

Comments pertaining to this report are invited and should be forwarded to: Director, Strategic Studies Institute, U.S. Army War College, 122 Forbes Ave., Carlisle, PA 17013-5244. Copies of this report may be obtained from the Publications and Production Office by calling commercial (717) 245-4133, FAX (717) 245-3820, or via the Internet at rummelr@awc.carlisle.army.mil

*****

Selected 1993, 1994, and all later Strategic Studies Institute (SSI) monographs are available on the SSI Homepage for electronic dissemination. SSI's Homepage address is: http://carlisle-www.army.mil/usassi/welcome.htm

*****

The Strategic Studies Institute publishes a monthly e-mail newsletter to update the national security community on the research of our analysts, recent and forthcoming publications, and upcoming conferences sponsored by the Institute. Each newsletter also provides a strategic commentary by one of our research analysts. If you are interested in receiving this newsletter, please let us know by e-mail at outreach@awc.carlisle.army.mil or by calling (717) 245-3133.

ISBN 1-58487-004-4

CONTENTS

Foreword v
1. Introduction 1
2. Foundations of Strategic Modernization 5
3. China's Quest for Information Dominance 25
4. Dawn of a New Age:
   China's Long-Range Precision Strike Capabilities ... 79
5. In Defense of its Own:
   China's Aerospace Defense 109
6. Conclusion 135
Appendix I. China Aerospace Corporation
   Organization 147
Appendix II. The Legend of Qian Xuesen 169
Appendix III. Space Support for Strategic
   Modernization 173
Appendix IV. China's Directed Energy Weapons 195
Appendix V. Commission of Science, Technology,
   and Industry for National Defense 215
Appendix VI. China's Ministry of Electronics
   Industry (MEI) 219
About the Author 229

FOREWORD

Conventional wisdom portrays the People's Republic of China (PRC) People's Liberation Army (PLA) as a backward continental force that will not pose a military challenge to its neighbors or to the United States well into the 21st century. PLA writings that demonstrate interest in exploiting the revolution in military affairs (RMA) are dismissed by a large segment of the PLA-watching community as wistful fantasies.

Major Mark A. Stokes, assistant air attaché in Beijing from 1992-1995, offers an alternative perspective. In this study, funded by the United States Air Force Institute for National Security Studies, he outlines emerging PLA operational concepts and a range of research and development projects that appear to have been heavily influenced by U.S. and Russian writings on the RMA. Fulfillment of the PLA's vision for the 21st century could have significant repercussions for U.S. interests in the Asia-Pacific region.

Major Stokes ventures into facets of PLA modernization that are often ignored. Backed by extensive documentation, he argues that the revolutionary modernization of the PRC's telecommunications infrastructure, a robust space-, air-, and ground-based sensor network, and prioritization of electronic attack systems could enable the PLA to gain information dominance in future armed conflicts around its periphery. Information dominance would be further boosted by China's traditional emphasis on information denial and deception.

In discussing the most likely scenario for PLA military action, Major Stokes postulates that information dominance—supported by a new generation of increasingly accurate and lethal theater missiles—could give the PLA a decisive edge in a future conflict in the Taiwan Strait. Highly accurate conventional theater missiles would play an especially critical role in rapid establishment of air superiority by suppressing airbases and neutralizing air defenses. Furthermore, the author argues that the PLA is striving to develop the capacity to complicate U.S. intervention in a Taiwan crisis.

In his appendices, the author provides an initial glimpse into PLA military space and directed energy weapons development. With extensive foreign technical assistance, China is investing in largely dual-use space-based systems that could provide the PLA with a valued-added boost to its overall military capabilities. In addition, China's development of "new concept" directed energy weapons— including high powered microwave and high powered lasers—may become a reality in the not-too-distant future.

Stokes argues that, while the PLA faces obstacles in fulfilling its modernization objectives, underestimating China's ability to make revolutionary breakthroughs in key areas could have significant ramifications for U.S. national security interests.

LARRY M. WORTZEL
Colonel, U.S. Army
Director, Strategic Studies Institute

CHAPTER 1

INTRODUCTION

Over the course of the last decade, the Chinese defense-industrial complex has initiated a focused strategic modernization program to meet the requirements of 21st century warfare. Chinese leaders, faced with numerous perceived national security challenges, have called for a readjustment of the People's Liberation Army (PLA) doctrine requiring the modernization of its space, information, long-range precision strike, and other strategic dimensions of warfare.

This study offers an alternative analytical approach to understanding PLA modernization. Ground force analysts, who tend to dominate the PLA studies field, naturally view the PLA mostly in terms of field equipment such as tanks, artillery, basic soldier skills, maneuver, and firepower. Air force analysts usually focus on aircraft, aircraft production, and pilot skills. Navy analysts think of the PLA in terms of destroyers, frigates, and other sea-based platforms. From these perspectives, the PLA is viewed as hopelessly backward, with little prospect, at least within the next 15-20 years, of becoming a challenge to any power in the region. These analysts are not necessarily wrong in their conclusions. Conventional PLA ground, air, and naval forces are woefully inadequate, and it is difficult to believe that they will be able to overcome shortcomings in the short to mid-term.

These traditional approaches, however, often gloss over strategic aspects of PLA modernization. There is an influential segment within China's defense-industrial complex which is concentrating on the development of doctrine and systems designed to enable targeting of adversarial strategic and operational centers of gravity, and defend its own, in order to pursue limited political objectives with an asymmetrical economy of force. In other words, the PLA, as part of its long-range regional security strategy, is attempting to develop an ability to target an enemy's forward-based command, control, communications, computers, and intelligence (C^4I) nodes, airbases, aircraft carriers and sea-based C^2 platforms, as well as critical nodes in space. Knowing the importance of protecting centers of gravity, Beijing is also prioritizing the fielding of systems to defend its own.

China's strategic modernization, if successful, will enable the PLA to conduct operations intended to directly achieve strategic effects by striking the enemy centers of gravity. These operations are meant to achieve their objectives without having to necessarily engage the adversary's fielded military forces in extended operations. Strategic attack objectives often include producing effects to demoralize the enemy's leadership, military forces, and population, thus affecting an adversary's capability to continue the conflict.

One of the key target systems of strategic attack is an adversary's command and control system. As Chinese and Western analysts have pointed out, disrupting the ability to communicate can be a critical step toward achieving strategic paralysis and disunity by cutting off the enemy's leadership from the civilian populace and fielded military forces. Conversely, protecting centers of gravity and the ability to command and control friendly forces are key objectives of strategic defense modernization.

There is no question the PLA is modernizing its air, naval, and ground forces across the board. However, since the mid-1980s, the PLA has placed special focus on certain enabling technologies which, short of resorting to weapons of mass destruction, would allow it to give play to its own strengths and exploit adversarial "Achilles heels." The timeframe for these capabilities varies. China's capabilities in ballistic missiles, ground-based radars, and information denial are already among the best in the world. Development of other competencies are much further down the road and are part of China's long-range plan beyond 2010.

This research project has numerous objectives, including the following:

- Determining the drivers behind China's strategic military modernization program;

- Identifing key research academies, institutes, and factories involved in strategic modernization;

- Tracing developments and programs related to ballistic and cruise missiles, directed energy weapons, military space program, and information warfare;

- Appraising the level of technology transfer into China which has contributed to its strategic modernization goals;

- Examining potential organizational and doctrinal changes within the PLA to related to new strategic systems;

- Assessing obstacles which may inhibit development and deployment of future strategic weapons systems; and,

- Analyzing ramifications of China's strategic modernization program on regional and U.S. national security.

To meet these objectives, this monograph is divided into four chapters following this introductory chapter. Chapter 2 contains an examination of the foundation for China's military modernization, to include drivers and defense research and development (R&D) strategy. Chapter 3 addresses one of the most critical, and often ignored, aspects of PLA modernization—its aspiration for information dominance. The monograph then turns, in Chapter 4, to an examination of systems which could enable the PLA to carry out paralyzing strikes against an adversary's critical nodes, drawing attention to R&D into increasingly lethal ballistic and cruise missiles. Chapter 5 addresses China's efforts to develop a comprehensive air and space defense network which brings into play many of the developmental efforts discussed previously. Chapter 6 discusses various obstacles to China's strategic modernization objectives and implications for the United States.

CHAPTER 2

FOUNDATIONS OF STRATEGIC MODERNIZATION

This chapter begins with an introduction of the key players within China's defense industrial complex, driving forces behind the strategic modernization program, and an overview of China's defense R&D strategy. An analysis of emerging PLA capabilities must begin with the technological base found within the defense industries. The defense industries operate through a mixture of bottom-up initiative from individual research institutes, and top-down direction from the Central Military Commission (CMC) and State Council. Emerging PLA doctrine, various threat perceptions, and bureaucratic competition are major driving forces for R&D of new weapon systems. Understanding approaches to R&D strategy, to include various developmental phases, provides a useful analytical approach to estimating timelines.

The Defense Industrial Complex.

The foundation of China's strategic modernization is its defense industrial complex. China has one of the largest and most cumbersome defense industrial establishments in the world. However, only certain segments are key players in China's strategic modernization. Overall responsibility lies with the CMC, which is roughly the counterpart of the U.S. National Command Authority. The CMC works closely with the State Council which conducts national economic planning, and provides policy guidance for general R&D and production. High-level policy related to priority functional R&D areas stems from various state leading groups.[1]

Central Special Committee. One other long-standing organization for policy decisions related to China's strategic modernization is the Politburo's Central Special Commission (CSC). The CSC was initially created to oversee China's strategic missile program in 1962. As of 1992, the CSC was directed by Li Peng, with Liu Huaqing as a deputy, and had as many as 20 members to coordinate, oversee, and make policy decisions at the highest level. Members include most vice premiers and leadership of the military industrial complex. Through the years, this organization has been sometimes referred to with different names, including National Defense Industry Special Commission, Sophisticated Weapons Production Commission, and Central Science and Technology Special Commission.[2]

State Science and Technology Commission, COSTIND, and the Defense Industries. The State Council's State Science and Technology Commission (SSTC) and the PLA's Commission of Science, Technology, and Industry for National Defense (COSTIND) oversee China's defense industries.[3] Directed by Song Jian, the SSTC is responsible for China's overall science and technology (S&T) development and manages the S&T budget, which has tripled over the last 2 years.[4] Key industries involved in strategic modernization include China Aerospace Corporation (CASC), China National Nuclear Corporation (CNNC), and Ministry of Electronics Industry (MEI). The general manager of each of these state conglomerates, in theory, reports to Song Jian for administration and civilian production, and to LTG Cao Gangchuan of COSTIND for military contracts.

China Aerospace Corporation. Among the defense industries, the CASC is the most important in China's strategic modernization effort.[5] CASC oversees the development of ballistic and cruise missiles, air defense systems, and the entire range of space launch vehicles and satellites. CASC is divided into five primary research academies (*yanjiuyuan*) which focus on launch vehicles and ballistic missiles (First Academy); air and missile defense systems, and developmental research into anti-satellite technology (Second Academy), anti-ship and land attack cruise missiles (Third Academy); solid propellant technology (Fourth Academy); and satellites (Fifth Academy). Another academy in Shanghai (Eighth Academy) augments and occasionally competes against the five basic academies in launch vehicles and air defense systems. A system of bases (*jidi*), numbered 061 through 068, located around the country have a wide range of manufacturing and R&D responsibilities. Originally established to provide Third Line manufacturing services subordinate to the individual academies, the bases have become independent R&D and manufacturing centers.[6]

Drivers of Strategic Modernization.

What drives Chinese strategic modernization? Is their modernization planning based on a rational top-down model which assesses fundamental values, interests, and national security requirements? Do they conduct scenario-based or threat-based planning? Are requirements subjected to bureaucratic influences between service arms and influence by industrial magnates? The answers to these questions are all probably yes. The mechanics of Chinese defense modernization planning are complex and opaque, probably not too different from the Pentagon. A number of factors, including emerging doctrine, threat perceptions, and bureaucratic politics, influence Chinese strategic planning, indigenous research and development, and acquisition.

Doctrine. The most important element driving China's military modernization is an emerging doctrine which emphasizes strategic attack against the most critical enemy targets. A major focus of China's military modernization is on gaining the ability to strike strategic targets, thus presenting the prospect of ending a conflict quickly by destroying an enemy's ability to wage war, or by convincing him to desist without first having to fight and defeat his military forces. Ultimately, the objective is to cause enough damage that the enemy will decide to heed Beijing's will or make it impossible to fight.

Strategic attacks, which are inherently offensive, often comprise the most direct means available to force an enemy to cease fighting or to make political decisions in line with Chinese objectives. They are executed to achieve maximum destruction of an adversary's ability to wage war. Such an approach to warfighting seeks to strike directly at the will and war-making capacity of the enemy unencumbered by surface military operations. As Major General Shen Xuezai, head of the Academy of Military Sciences' (AMS) Military Systems Department, states, "Only by controlling the entire battlespace and striking at key points so as to paralyze the enemy's entire operational system and immobilize its forces, will it be possible to win a war."[7] Strategic attacks rely on forcing the enemy into the reactive and defensive mode, allow retention of the initiative, and reduce threats to friendly units. Strategic campaigns, of course, are only part of an overall war effort and only one means available to those in power who wield national military instruments. Widespread destruction of enemy infrastructure and fielded forces is not necessary.

Drawing doctrinal lessons from U.S. conduct of the Gulf War, many Chinese strategists view the most critical targets as the national command and control apparatus, to include leadership, operational command centers, and the C^4I infrastructure. Other classes of targets include key manufacturing, petroleum storage, and power generation facilities; transportation infrastructure; population centers; and fielded military forces. Attacking fielded military forces, especially reconnaissance and strike assets, may be necessary as more of a defensive measure to protect friendly centers of gravity.

This strategic attack doctrine, one aspect of the PLA's "limited war under high tech conditions" (*jubu zhanzheng zai gaojishu tiaojian xia*), has a number of characteristics. First, the PLA's emerging doctrine continues to adhere to the traditional strategy of "pitting the inferior against the superior" (*yilie shengyou*), which recognizes technological inferiority for an indefinite period of time. Relative superiority in selected areas, however, can be applied against an enemy's weaknesses. The emerging doctrine is offensive in nature. Chinese doctrinal literature indicates that, if war against a technologically superior power breaks out, an enemy will likely deploy forces rapidly and then launch a massive air campaign. While an enemy is amassing forces, a window of opportunity exists for a preemptive strike.[8]

From the Chinese perspective, this asymmetrical approach, "gaining the initiative by striking first" (*xianfa zhiren*), is an effective means to offset technological and logistical advantages which a more advanced military power brings to the fight. This approach to warfare marks a significant change from the previous principle of "gaining mastery only after the enemy has struck" (*houfa zhiren*). The emerging doctrine requires a high degree of secrecy, mobility, a highly accurate concentration of firepower, and surprise. These principles make possible a quick resolution of conflict. This does not require annihilation of the enemy or occupation of his territory, only a paralyzing "mortal blow" (*zhiming daji*), "winning victory with one strike" (*yizhan, ersheng*).[9]

Other important aspects of PLA's doctrinal shift is recognition of a multidimensional, greatly expanded battlespace, encompassing the electromagnetic spectrum, land, air, sea, and space. Only a small proportion of the PLA's total troop strength is necessary to achieve their military objectives. This principle, "winning victory through elite troops" (*jingbing zhisheng*), requires only a small number of highly educated troops in order to achieve a quick victory. The concept of a swift victory is another principle, "fighting a quick battle to force a quick resolution" (*suzhan sujue*). Another principle is priority on the "in-depth strike" (*zongshen daji*) to hit distant, strategic targets in order to avoid forward troops and hit vital centers of gravity, such as ports, critical C^4I nodes, airbases, and supply centers. Cruise and ballistic missiles, and other long-range precision strike assets, are crucial in conducting these in-depth strikes.[10]

PLA writings strongly indicate the foundation of the emerging doctrine is the concept of information dominance (*zhixinxiquan*). Information dominance is generally achieved through command and control warfare, using combinations of airpower, special forces, and strategic missile units to strike an adversary's C^4I infrastructure. Interfering with an enemy's ability to obtain, process, transmit, and use information can paralyze his entire operational system.[11]

Although dissenting opinions probably exist, there are clear indications that emerging PLA doctrine is pushing the development of long-range precision strike and information operations. Because of this, the CMC, through COSTIND, has granted priority to offensive and defensive information systems and to long-range precision strike assets, primarily in the form of ballistic and cruise missiles. At the same time, China is emphasizing development of defensive systems intended to counter enemy information operations and long-range precision strikes against Chinese territory.

U.S. and Russian Missile Defense Programs. Since the late 1970s, another key driver for China's strategic modernization has been U.S. and Russian efforts to develop missile defense systems. Preliminary work on the U.S. strategic missile defense program began in the late 1970s and emerged as the Strategic Defense Initiative (SDI) in March 1983. Since then, offshoots of SDI have included Global Protection Against Limited Strike (GPALS), and various National and Theater Missile Defense programs. Current and future national missile defense concepts, including theater high-altitude area defense systems (THAADS) and the experimental Space-Based Laser could be deployed as early as the turn of the century.

Since the late 1970s, the CMC, COSTIND, the Second Artillery, and the space and missile industry have been extremely concerned about the deployment of U.S. and Russian missile defense systems. COSTIND and the space and missile industry have been especially active in keeping appraised of technological and political developments concerning missile defense and have been developing technologies which will assure China a second strike capability. Maintenance of a viable nuclear deterrent has been the highest priority for Chinese defense planners since the mid-to-late 1950s. This is still true today.

This fear of losing their retaliatory capability began shortly after the announcement of SDI in 1983, when Chinese analysts began evaluating potential responses. The first important paper on SDI and its potential implications for China was prepared by the Ministry of Foreign Affairs (MFA) for Premier Zhao Ziyang in 1984. In late 1984 or early 1985, the central leadership tasked several ministries and research institutes to thoroughly examine the SDI issue and its implications for China. During 1985, the defense industrial complex hosted several conferences on SDI. A primary conclusion was that Soviet and U.S. development of ballistic missile defense (BMD) systems had significant implications for China's nuclear deterrent. By 1986, Chinese experts generally agreed there were three potential responses: expansion of offensive forces; development of countermeasures, such as shielding and spinning of ballistic missiles to penetrate BMD systems; and deployment of anti-satellite (ASAT) weapons to destroy space-based BMD systems.[12]

COSTIND played a key role in developing the Chinese response to the "global technical revolution" sparked by SDI. As the MFA was developing its SDI response, COSTIND held a March 1984 meeting on "How to Meet the Global Technical Revolution Challenge." In September 1984, COSTIND delivered a proposal to the CMC suggesting relevant PLA branches develop defense S&T gameplans out to the year 2000. Working in conjunction with the State Council, COSTIND formulated a defense S&T strategy which focused attention on certain key technologies and presented it at a November 1985 meeting with CMC leadership. Afterwards, in February 1986, COSTIND, with CMC support, sanctioned the overall long-term development effort and further directed the formation of 18 study groups to focus on designated critical technologies.[13]

However, some within the defense S&T community believed COSTIND's plan was not sufficient to meet the technical challenges posed by SDI. In March 1986, four of China's most prominent defense engineers presented a petition to the Central Committee on establishing a "High Technology Research and Development Plan Outline." All of the engineers pushing the new initiative were involved in strategic programs—Wang Daheng, a preeminent optics expert who played a role in China's space tracking network; Wang Ganchang, one of the founding fathers of China's nuclear program; Yang Jiachi, a satellite attitude control expert; and Chen Fangyun, an electronics engineer and leader of the program to develop China's space tracking network. The plan, referred to as the 863 Program, was implemented in parallel to COSTIND's Long Range Plan to Year 2000 and was jointly managed by COSTIND and the SSTC. The 863 Program, still a guide and funding source for numerous preliminary R&D projects, focuses on some of the same technologies included in the SDI and Europe's answer to SDI, the Eureka Program, including space systems, high powered lasers, microelectronics, and automated control systems.[14]

China's Gulf War Syndrome. Yet another driver for China's strategic modernization is U.S. performance in the Gulf War. The 1991 Gulf War was a rude awakening for the CMC and the military-industrial complex. The awesome display of military power showed the Chinese leadership just how vulnerable their homeland is to attack from a potential enemy. A second important conclusion which the CMC drew from the Gulf War was the preeminence of air power and long-range precision strike, augmented by information-based warfare, in a greatly expanded battlespace. According to one source, in a December 1995 meeting the CMC concluded "ground fighting can only enhance the results of battle." Key targets in warfare are the enemy's "nervous system and brain," rather than its ground targets and fighting units. Another important conclusion was the primacy of offensive operations as an essential element of defense.[15]

China's Revolution in Military Affairs (RMA). Lessons drawn from the U.S. experience in the Gulf War are being augmented by subsequent literature on the potential RMA. The increased doctrinal emphasis on air power, long-range precision strike, information warfare, and paralyzing strikes on centers of gravity has been further reinforced by calls to grasp concepts associated with the RMA. A number of influential strategists and high ranking members of the defense industrial complex have acknowledged the RMA and called on the PLA leadership to meet its challenges of 21st century warfare. Among the most influential are Zhu Guangya, COSTIND S&T Committee Chairman, and Qian Xuesen, Senior COSTIND Advisor. They advocate placing priority on understanding the nature of future warfare, investing in key enabling technologies such as microelectronics and information systems, raising the level of PLA officer technical understanding, and adopting innovative doctrine and organizational changes.[16] Chinese commentators view the RMA as driving four spheres of warfighting, including information warfare, precision strike, strategic maneuver, and space combat.[17]

Quest for Great Power Status. Yet another driver is the quest for status and procurement of weapon systems which earmark a country as a great power. From the Chinese perspective, there are certain technologies and weapons systems which a great power is simply expected to possess. The mindset is "whatever the West can do, we can do as well." The resulting technologies or systems, however, may not necessarily be optimal for China's defense needs. For example, China's development of an aircraft carrier could be useful against an regional adversary, but would be of little direct consequence to United States. Its real effect is to show that China is a great power.

Territorial Defense. China does take a logical and rational approach to meeting its defense requirements, to include enforcement of sovereignty claims. Assessments are made as to what the threats are that China will face in the near, medium, and long term, and how China can assure the capability to defend against those threats. PLA strategists believe China may be faced with surprise attacks from the East and South China Seas early in the next century. In order to guarantee territorial defense, Chinese policymakers have outlined a defense in-depth approach to national security and have adopted a strategy to ensure dominance within the seas area bounded by the first island chain (*diyi daolian*), bounded by Japanese home islands, the Ryuku Islands, Taiwan, the Philippines, and Borneo. The PLA has also begun to examine a territorial defense strategy that extends out to the second island chain (*dier daolian*), including the Marianas, Guam, and the Carolines.[18] The PLA Navy, Air Force, and Second Artillery figure largely in this defense strategy.

Territorial defense also includes sovereignty claims to Taiwan, the South China Sea, and the Diaoyu Islands in the East China Sea. Developing an ability to militarily subdue Taiwan is an especially important driver in China's strategic modernization. PLA planners, however, fully understand the United States maintains the option to intervene if China takes military action against Taiwan. From the PLA's perspective, American intervention in what Beijing believes to be an internal problem serves as the most likely scenario for a collision between China and the United States. Therefore, gaining the ability to defeat a superior power, like the United States, in a clash over Taiwan is a critical factor in China's force planning.

Support for the National Economy. Strategic modernization requirements must compete, or at least support, China's overall economic development. China's overarching objective is economic development, and fostering of an environment conducive to their economic security. The PLA plays two roles in supporting the national economy. First, development of systems may be emphasized due to their dual-use nature. COSTIND may be inclined to select and design systems which can produce some economic benefit. Information systems, satellites, and cruise missiles, for example, are systems which can either support the national economy or be sold abroad for profit. Projects which can be plausibly wrapped in a civilian facade make it much easier to import high technology items from countries which ban the sale of military items. Strategic modernization will become increasingly important to ensure China's lines of communication are open to the free flow of oil from the Middle East and other sources.

Organizational and Bureaucratic Politics within the Defense-Industrial Complex. The drive for organizational influence and power are as prevalent in China as they are in other countries. There is no doubt various PLA branches and services compete for finite budgets and resources, probably with the Second Artillery, Navy, and Air Force coming out on top. As China's missile defense program takes shape (discussed in more detail later), for example, the PLA Air

Force, Second Artillery, and the ground forces will likely compete for these crucial defense assets. Within the defense industries, competition for contracts, prestige, and profits between and within the space, electronics, aviation, and nuclear industries heavily influence program priorities and investment. The space industry, for example, lobbies for advancement of ground-based air defense systems over the aviation industry's fighter component. Competition between the aviation industry's Nanchang Aircraft Factory and CASC's Third Academy in development of cruise missiles is well-known. Personal connections between industrial leaders and the military and civilian leadership are also important. Examples include electronics industry cliques involving Jiang Zemin, Zou Jiahua, and Hu Qili; and space and missile industry cliques involving Bo Yibo, Liu Huaqing, Zhang Aiping, Ding Henggao, Qian Xuesen, and Liu Jiyuan.

Technological Advances. A final driver for strategic modernization is the seemingly endless stream of critical technologies flowing into China. Strategic programs put on hold in the 1970s and 1980s due to technical difficulties have been resurrected due to increased access to foreign technology and expertise. Since 1991, a massive flood of technology has significantly opened the realm of possibilities, and has influenced program decisions which are based on feasibility. First, the breakup of the former Soviet Union in 1991 and dire economic state of former Soviet defense industries have resulted in a shopping spree for China's research institutes. Russian and Ukrainian guest lecturers and annual technical exchanges assist Chinese engineers in solving difficult technical problems. The 1994 demise of COCOM and the lack of an effective successor regime have loosened up the flow of manufacturing, electronics, and materials technology.

A Guiding Light: Defense R&D Strategy.

There is little evidence to suggest that an overarching strategic planning document reflects these disparate influences and drivers and serves as the bible for long-range R&D and acquisition. There are, however, a handful of interrelated defense S&T plans which guide the efforts of China's defense industry.

The 863 Program. Much of China's strategic modernization effort is framed within a large scale plan designed to establish a technological foundation for the 21st century. High-level discussions on this strategic modernization plan, eventually called the 863 Program, were initiated shortly after the announcement of the SDI in March 1983 and the initiation of Europe's Eureka Project in 1985. The 863 Program is roughly comparable to the Eureka Program, which combined civilian and strategic defense programs.[20] The 863 Program, managed by COSTIND and the State Science and Technology Commission, was established as a mechanism to concentrate China's S&T establishment onto seven key areas for long-term development.[21] These areas include space, lasers, automation, biotechnology, information systems, energy, and new materials. Within these seven areas, 17 major projects were designated. Projects under the 863 Program are budgeted independently from the PLA, COSTIND, and SSTC budgets. COSTIND oversees 863 laser and space programs, while the SSTC manages the other five areas.[22]

China's defense R&D strategy has its roots in a directive laid out by Nie Rongzhen in the 1960s. The strategy, called "Three Moves in a Chess Game" (*sanbuqi*), calls for three models in the R&D cycle at any one time. Generally, this is defined as having a model in trial manufacturing and testing; a follow-on model under design; and preliminary research on basic technologies associated with a generation-after-next model.[23] The R&D cycle consists of the three broad phases of preliminary research, model R&D, and production.

Preliminary Research (yuxian yanjiu, or yuyan). Chinese defense industries stress on advance research serves as the foundation for follow-on stages of development. A solid foundation can significantly shorten design time—for example, one medium-range ballistic missile system took only 21 months to design, quite a feat compared to other systems which have taken up to a decade to design. Short R&D time is due to mastering of mature technologies through preliminary research. Preliminary research has two categories:

- Generic technologies applicable to multiple systems, i.e., telemetry, aerodynamics, GPS exploitation, hypersonics, artificial intelligence, etc;

- Technology applicable to a specific system, for instance, a movable spot beam antenna for a COMSAT or a new propulsion system for a missile.

Preliminary research, an integral part of a long-range plan, is carried out on a contract basis and strives to achieve technological breakthroughs. China has focused on key technologies in long-range plans, best exemplified by the 863 Plan. At the end of the allocated time for the project, approximately 45 percent of advance research projects move on to the following stage, model R&D, while another 40 percent are used as a foundation for follow-on advance research projects. Only 14 percent of preliminary research projects are civilian in nature.[24]

Model R&D (xinghao yanzhi). After completion of an preliminary research project, COSTIND may contract with a PLA service branch for further research and development, weaponization, and eventual deployment. This phase, model R&D, is lengthy, costly, and is closely monitored by COSTIND. Each model R&D program which emerges from the preliminary research phase requires a new COSTIND managed contract.[25] The model R&D phase is divided into four sub-phases:

- *General systems design.* Chief designer appointed who monitors various subsystem design efforts. Assessments of various design options are selected based on modeling and simulation.

- *Prototype.* Design is revised after a series of ground tests to ensure the model meets technical specifications.

- *Flight model.* After ground testing, flight tests are carried out beginning with simple ones, and traversing through a series of increasingly complex tests.

- *Commission.* After successfully completing flight testing, the system will go through a design certification board, and if approved, will enter small batch production.[26]

Chief Designer System and Dual Command. A chief designer is appointed at the initiation of the systems design phase to oversee subordinate subsystems design. Subordinate designers are often not within his research organization chain of command. The chief designer coordinates efforts among various research institutes, academies, academic centers, and industries. This chief designer organization is paralleled by an administrative chain of command, represented by industry, academy, and research institute leadership, which ensures requirements and timeliness standards of the PLA are being met. Administrative leadership oversees several programs at once along with their day-to-day management responsibilities. The administrative and technical sides work closely together to ensure an economy of effort, timeliness standards, manpower, materials, funding, and other considerations are attended to. The office of the chief designer is usually located within the systems design department of CASC or the individual academy, with the parallel administration oversight normally residing in the office of one of the deputy directors of CASC or academy. For especially complex projects which significantly draw from resources of various industries, COSTIND plays a leading administrative, and possibly even technical role.

Systems Engineering Design Departments. The systems engineering design departments of COSTIND, CASC, and the various academies and bases have a critical role in the development of future weapons systems and projects. Chief engineers work out of these organizations which have a variety of functions. First, with input from subordinate information institutes[27] and other entities, they offer systems development recommendations to industry leadership, COSTIND, State Council, and CMC, and on systems development issues. They analyze technical options and, as industry planners, provide influential recommendations on national level mid- to long-range plans and developmental objectives. The departments also support individual chief designer offices within the department. In addition, systems design departments are responsible for program evaluations and reviews and for overall quality control. The following list exemplifies organizations responsible for systems engineering:[28]

- Beijing Institute of Systems Engineering (COSTIND);

- China Academy of Electronics and Information Technology, Systems Engineering Department (MEI-level);

- China Institute of Astronautical Systems Engineering (CASC-level);

- First Planning Department (CASC's First Academy);

- Beijing Institute of Electronic Systems Engineering (CASC's Second Academy);

- Beijing Institute of Electromechnical Systems Engineering (CASC's Third Academy);

- 41st Research Institute (CASC's Fourth Academy);

- 501st Research Institute (CASC's Fifth Academy);

- Shanghai Institute of Systems Engineering (CASC's Shanghai Academy of Space Technology).

ENDNOTES - CHAPTER 2

1. There are at least seven state leading groups which provide interagency policy guidance for such functional areas as telecommunications, space, and lasers.

2. John Wilson Lewis and Xue Litai, *China Strategic Seapower: The Politics of Force Modernization in the Nuclear Age*, Stanford: Stanford University Press, 1994, pp. 78-79, 262 (footnote #43), 277 (footnote #30).

3. At the time of this writing, SSTC was being transformed into the Ministry of Science and Technology. COSTIND's military functions are being placed under the purview of the PLA General Armaments Department (*zongzhuangbei bu*), equal in stature to the General Staff Department. Civilian aspects of the old COSTIND structure have been melded into a new COSTIND under the State Council.

4. The R&D budget has been increased to 1.5 percent of GDP, still relatively low, as compared to the United States and Japan (both 2.72 percent of GDP in 1993). See Norman Wingrove, "China Sees Tripling of R&D Spending As Key to 21st Century Economic Power," *Research Technology Management*, November-December 1995, pp. 2-3.

5. At the time of this writing, CASC was being subordinated under the newly restructured civilian Commission of Science, Technology, and Industry for National Defense (COSTIND).

6. See Appendix I for a detailed outline of the CASC organizational structure. Since this study was completed, CASC has been divided into two entities: China Aerospace S&T Corporation and China Aerospace Electro-mechanical Corporation. The S&T Corporation has adopted CALT, the Fourth Academy, CAST, 062 Base, and 067 Base. The Electro-mechanical Corporation has integrated the Second and Third Academies and the remaining bases.

7. Major General Shen Xuezai, "The New Military Revolution and Change in Military Organizational Structure," *Zhongguo Junshi Kexue* (*China Military Science*), February 20, 1998, pp. 122-130, in *Foreign Broadcast Information Service, China* (henceforth *FBIS-CHI*)-98-167. Major General Shen, as director of the AMS Military Systems Department, is one of the PLA's most influential figures in the development of strategy and operational doctrine.

8. Senior Colonel Jiang Lei, *Xiandai Yilie Shengyou Zhanlue (Modern Strategy of Pitting the Inferior Against the Superiority)*, Beijing: National Defense University Press, pp. 6-49. Colonel Jiang is

one of the few PLA officers awarded a Ph.D. in Operations Research from AMS. He is currently assigned to AMS Strategic Studies Department. On the preemptive strike concept, see Lu Linzhi, "Pre-emptive Strikes Endorsed for Limited High Tech War," *Jiefangjun Bao*, February 14, 1996, in *FBIS-CHI*-96-025.

9. For one of the best overviews of these doctrinal shifts, see Nan Li, "The PLA's Evolving Warfighting Doctrine, Strategy and Tactics, 1985-95: A Chinese Perspective," in *China Quarterly*, July 1995, pp. 443-463. For more detail, see Liu Mingtao and Yang Chengjun, *Gaojishu Zhanzhengzhong de Daodanzhan* (*Missile War Under High-Tech Conditions*), Beijing: National Defense University Press, 1993, pp. 5-26; also see Li Qingshan, *Xin Junshi Geming Yu Gaojishu Zhanzheng* (*New Military Revolution and High Tech Warfare*), Beijing: AMS Press, 1995; Liu Senshan and Jiang Fangran, *Gaojishu Jubu Zhanzheng Tiaojianxia de Zuozhan* (*Operations Under High Tech Local War Conditions*), Beijing: AMS Press, 1994, pp. 13-33.; and Senior Colonels Huang Xing and Zuo Quandian, "Operational Doctrine for High Tech Conditions," *Zhongguo Junshi Kexue* (*China Military Science*), November 20, 1996, pp. 49-56, in *FBIS-CHI*-97-114.

10. *Ibid.*

11. Among numerous references, see, for example, Wang Pufeng, *Xinxi Zhanzheng yu Junshi Geming* (*Information Warfare and the Military Revolution*), Beijing: AMS Press, 1995.

12. Bonnie S. Glaser and Banning N. Garrett, "Chinese Perspectives on the Strategic Defense Initiative, *Problems of Communism*, March-April 1986, pp. 28-44.

13. *China Today: Defense Science and Technology*, Beijing: National Defense Industry Press, 1993, pp. 149-150.

14. *Ibid*, pp. 152-153; also see Richard P. Suttmeier, "China's High Technology: Programs, Problems, and Prospects," *China's Economic Dilemma*, pp. 546-564.

15. Nan Shih-yin, "Inside Story of Enlarged Central Committee Meeting," Hong Kong *Kuang Chiao Ching*, January 16, 1996, in *FBIS-CHI*-96-027; also see Jen Hui-wen, "Latest Trends in China's Military Revolution," Hong Kong *Hsin Pao* (*Hong Kong Economic Journal*), February 9, 1996, in *FBIS-CHI*-96-047. For other comments on lessons from the Gulf War, see Ho Po-shih, "The Chinese Military Is Worried About Lagging Behind in Armament," *Tangdai*, March 9, 1991, pp. 17-18.

16. See Zhu Guangya, "Current Engineering Technology and the New Revolution in Military Affairs," *Zhongguo Junshi Kexue (China Military Science)*, February 20, 1996, in *FBIS-CHI*-96-246.

17. Liang Zhenxing, "New Military Revolution and Information Warfare," *Zhongguo Dianzi Bao (China Electronic News)*, October 24, 1997, p. 8, in *FBIS-CHI*-98-012.

18. Lewis and Xue, pp. 229-230.

19. For a theoretical perspective on organizational and bureaucratic models of policy making, see Graham T. Allison, *Essence of Decision: Explaining the Cuban Missile Crisis*, Boston: Little, Brown, and Company, 1971.

20. See "China's Spectacular Plan for tackling Key Scientific and Technological Problems," *Zhongguo Xinwen She*, May 21, 1986, in *FBIS-CHI*, May 30, 1986, p. K17.

21. "*Space*" in Chinese, "*hangtian*," refers to both space systems as well as ballistic, cruise, and surface-to-air missiles.

22. *China Today: Defense Science and Technology*, pp. 151-153. The 863 Program was sparked by a petition to the Central Committee issued by four prominent Chinese scientists, Wang Daheng, an optics expert; Wang Ganchang, one of China's leading nuclear weapons engineers; Yang Jiachi, a control theory expert; and Chen Fangyun, an electronics expert.

23. An example of this approach would be development of the 1,800 kilometer range DF-21 ballistic missile. As this system was being tested and deployed in Second Artillery experimental regiments in the mid-1980s and early 1990s, designers within CASC's First Academy were already designing an upgraded variant of the DF-21, the DF-21A. At the same time, research institutes were examining more advanced technologies which could be integrated into a follow-on for the DF-21A.

24. *China Today: Defense Science and Technology*, pp 155-156.

25. *Ibid.*

26. R&D phases vary between industries and academies. For example, the First Academy divides its phases of missile model R&D as follows: planning phase (*fang'an*); initial prototype (*chuyang*); testing prototype (*shiyang*); design finalization (*dingxing*). See Gan Chuxiong and Liu Jixiang, *Dandao Daodan Yu Yunzai Huojian Zongti Sheji*

(General Design of Missiles and Launch Vehicles), Beijing: National Defense Industry Press, January 1996, p. 42.

27. Each academy has an institute which is responsible for research on foreign technology and assessing applications for Chinese systems. In a sense, in addition to COSTIND and CASC headquarters, each academy has its own intelligence apparatus unique to its own specialty. These institutes also serve as advocates for their projects. For example, the Second Academy's 208th Research Institute publishes technical papers on the U.S. THAAD and KKV ASAT programs for use by their academy's engineers. These types of technical papers are abstracted in *China Astronautics and Missilery Abstracts (CAMA)* and other journals. The 208th also is CASC's mouthpiece for advocating further investment into ground based air defense systems.

28. Information from *China Today: Space Industry*, Beijing: Astronautic Publishing House, 1992, pp. 430-450.

CHAPTER 3

CHINA'S QUEST FOR INFORMATION DOMINANCE

Beijing's highest priority for strategic modernization is in the realm of information. Competition for information is not a new concept in China. A review of Sun Tzu's *Art of War* reflects the Middle Kingdom's traditional emphasis on information dominance. For centuries, Chinese leaders have been masters at collecting, controlling, and manipulating information. Building upon this traditional core competency, Beijing is aggressively absorbing technologies associated with the on-going information revolution. In fact, the international business community views the China market as the largest in world and is devoting considerable attention to meeting the Chinese demand for technology. China's unparalleled revolution in information technology will provide the foundation for its overall strategic military modernization. In fact, China's information revolution is changing the fundamental character of the PLA, perhaps much more rapidly than most observers appreciate.

Since the conclusion of the Gulf War, a growing chorus of PLA officials have strongly advocated the aggressive pursuit of information-based warfare doctrine and systems. Analysts claim the PLA first addressed information warfare earlier than the United States or Russia, when, in 1985, a PLA field officer named Shen Weiguang wrote a book entitled *Information Warfare*, excerpts of which were published in the PLA's leading newspaper, *Jiefangjun Bao*, in 1987. Since that time, numerous PLA think-tanks and COSTIND organizations have published articles and held symposiums on information warfare. Some of China's most influential figures, including Qian Xuesen and Zhu Guangya, have called for full-scale adoption of information-based warfare as the foundation for China's revolution in military affairs. U.S. conduct of the Gulf War impressed a growing number of information warfare advocates, such as Qian Xuesen, who, in the Third Annual COSTIND S&T Committee meeting in March 1994, called for establishment of a national information network and associated technologies.[1]

COSTIND sponsored a symposium, "Analysis of the National Defense System and the Military Technological Revolution" in December 1994 and followed up with another conference, "The Issue of Military Revolution," in October 1995. A number of entities have combined to form an informational warfare research institute, and there are indications China will form an information warfare simulation center.[2]

Enthusiasm reached a crescendo when China's leading authorities on strategy and warfare convened in Shijiazhuang for a "Forum for Experts on Meeting the Challenges of the World Military Revolution" in December 1995. The consensus of the group was to meet the challenges of the information-driven revolution in military affairs. The more than 30 high-ranking experts attending the conference called for the development of weapons which can "throw the financial systems and army command systems of the hegemonists into chaos." These types of weapons are useful for underdeveloped countries to use against a nation which is "extremely fragile and vulnerable when it fulfills the process of networking and then relies entirely on electronic computers." China must abandon the strategy of "catching up" with more advanced powers and "proceed from the brand new information warfare and develop our unique technologies and skills, rather than inlay the old framework with new technologies." Some observers believe by adopting information-based approaches to warfare, China can effectively leapfrog into the 21st century as a preeminent military power.[3]

PLA emphasis on information dominance was further reinforced when, at a December 1995 COSTIND National Directors meeting, Vice Chairman of the Central Military Commission General Liu Huaqing allegedly stated:

> Information warfare and electronic warfare are of key importance, while fighting on the ground can only serve to

exploit the victory. Hence, China is more convinced (than ever) that as far as the PLA is concerned, a military revolution with information warfare as the core has reached the stage where efforts must be made to *catch up with and overtake rivals.* (emphasis added)[4]

There is a growing chorus of opinion propelling China toward establishing a national strategy for developing a comprehensive information warfare capability. Components of this strategy include educating China's domestic audience and the PLA on the role of information in warfare through frequent articles in the PLA's newspaper, *Jiefangjun Bao*, and in academic journals, such as the AMS' *Zhongguo Junshi Kexue (China Military Science).* There is also a push toward establishment of a high-level leading group on national defense information modernization which can study and draft information warfare theory applicable to China's unique defense requirements, and formulate plans and programs for China's defense information infrastructure. Chinese observers stress the need to develop "perfect weapons" which serve as "trump cards" (*shashoujian*) to exploit an adversary's reliance on sophisticated microelectronics.[5] Other aspects of developmental strategy include focusing efforts to achieve technical breakthroughs in information technology and stepping up the education and training of PLA soldiers, airmen, and sailors on information warfare-related skills.[6]

This discussion on China's quest for information dominance first examines the most critical aspect of China's information revolution—development of its microelectronics capability. The next section addresses Beijing's increasing capability to field sophisticated ground, air and space-based sensors. Subsequent sections examine China's development of its national information infrastructure, Beijing's information control programs, and China's evolving information attack capabilities.

Information Warfare Concepts and Doctrine.

The PLA strategic community is working furiously to develop information warfare doctrine and concepts. Over the last few years, the AMS and the National Defense University (NDU), two of the PLA's leading think-tanks, and COSTIND have flooded China's strategy community with writings on information warfare. Chinese commentators define information warfare in broad terms. For example, COSTIND defines information warfare as follows:

> All types of warfighting activities that involve the exploitation, alteration, and paralysis of the enemy's information and information systems, as well as those activities which involve protecting one's own information and information systems from exploitation, alteration, and paralysis by the enemy.[7]

There are several common principles included in discussions of China's approach to information warfare. First and foremost is the concept of information dominance (*zhixinxiquan*), described as the ability to defend ones own information while exploiting and assaulting an opponent's information structure. As some advocates have pointed out,

> The key to gaining the upper hand on the battlefield is no longer mainly dependent on who has the stronger firepower, but instead depends on which side discovers its enemy first, responds faster than the latter, and strikes more precisely than the latter.[8]

Information dominance is generally achieved through command and control warfare, using combinations of airpower, special forces, and strategic missile units to strike an adversary's C^4I infrastructure. Interfering with an enemy's ability to obtain, process, transmit, and use information can paralyze his entire operational system. Crucial to achieving the paralysis effect is increased analysis of the enemy military infrastructure as a system.[9] The multiplying effect of intelligence and information leads to an almost limitless expansion of combat effectiveness. According to some Chinese observers, warfare in the information era is a test of strength between intelligence capabilities of combat forces. Information superiority, say Chinese commentators, is not necessarily determined by technological superiority, but by new tactics and independent creativity of commanders in the field. In fact, Chinese operational units have begun to integrate information warfare concepts into their training.[10]

Another concept is integration (*yitihua*) and seamless operations (*feixianxing zuozhan*), defined as tying together the five dimensions of warfare—air, land, sea, space, and the electromagnetic spectrum, integrating sensors with mobile missiles, air, and sea-based forces. Chinese information warfare analysts view the battlespace as having been greatly expanded with no clear battlelines. Operational emphasis is on deep strike (*zongshen zuozhan*) against enemy command hubs, information processing centers, and supply systems. Sudden and quick (*turanxing yu kuaisuxing*) strikes at the "vital point" (*dianxue*) of the enemy's information and support systems can paralyze the enemy and destroy his morale.

Operational objectives are not the seizing of territory nor the killing of enemy military personnel, but rather the destruction of the other side's willingness or capability to resist. Over-the-horizon warfare (*yuanzhan*) has replaced close-in fighting, and from the Chinese perspective, will become the main strike force of the future. Chinese information warfare specialists advocate exploiting enemy reliance on complex computers. By destroying an enemy's computer systems, high-tech weaponry can no longer operate. Information warfare specialists also highlight cognitive aspects of information warfare. One concept outlines a five-part approach to offensive information warfare, emphasizing targeting adversarial cognitive systems, jamming, and information manipulation.[11]

The Foundation of Its Information Revolution: China's Development of Microelectronics.

There is an intimate relationship between information dominance and microelectronics. Recognizing this, since 1991 China has named the development of indigenous microelectronics as a top national security priority. Dual-use microelectronics and telecommunications equipment support both economic development and military modernization. China's unparalleled revolution in information technology—

to include enabling technologies such as microelectronics, computer systems and software, and artificial intelligence—will provide the foundation for its overall strategic military modernization. Over the past decade, information technology has been the fastest growing segment of China's economy, rising at an annual rate of 30 percent.[12]

Among these enabling technologies, the Chinese defense industrial complex most recognizes the central role of microelectronics. Priority investments in their integrated circuit R&D and production capability could translate into significant payoffs in the next few years. For example, in 1993 China's space and missile industry formed a research academy—equal in stature to those academies which develop cruise missiles and satellites—dedicated toward the development of space-qualified microelectronics. Since the lifting of international export controls on microelectronics in the last few years, the use of foreign integrated circuit (IC) production technology, to include U.S., Japanese, and Korean, is becoming more widespread.

A solid foundation in microelectronics is essential for Beijing's strategic modernization goals. At the heart of China's indigenously developed information systems and guidance platforms is the IC. The Chinese defense establishment recognizes IC chips as the most basic component for their long-term economic health and national strength. In 1993, China depended on foreign sources for up to 80 percent of its ICs. However, by the end of the 8th Five-Year plan in 1995, China was able to meet half of its total IC demand with domestically produced ICs. Over the last few years, China's production of integrated circuits has increased at a rate greater than 20 percent per year. As a highlighted area under the 863 Program, microelectronics has become a strategic industry. The primary focus is on reducing the cost and weight of microelectronics while increasing reliability and survivability.[13] Increasing emphasis is being placed on development of very large scale integrated circuits (VLSIC) which have applications in advanced phased array radars and space systems.[14]

The leading organization for microelectronics is the MEI.[15] Formerly known as the 10th Ministry of Machine Building Industry, MEI carries out R&D on microelectronics, telecommunications, radars, electronic warfare systems, computers, and systems integration software. Directed by political heavyweight Hu Qili, MEI has entered into a number of joint ventures as a means to raise its technical competence.[16] Joint venture arrangements often include the transfer of IC production lines. Shougang-NEC Electronic Company produces 5000 chips a month and is expected to average 8000 chips in 1997. The production line is not yet up to mass producing sub-micron ICs, but is moving in that direction.[17] One Sino-Japanese joint venture producing sub-micron (0.35 micron) integrated circuits is located in the Shangdi Information Industry Base in northwest Beijing.[18]

Moving beyond sub-micron integrated circuits, the Chinese defense industrial complex is placing more emphasis on microelectromechanical system (MEMS) technology. MEMS is vital for miniaturizing components to reduce weight and volume of military systems, to include satellites, robotic reconnaissance systems, and nano-technological weapon systems. MEMS applications also include miniature inertial measurement units for brilliant munitions; merging information processing with sensors; and for miniature, low cost deformable mirrors to compensate for atmospheric distortions when using high-powered lasers. There is a movement, driven in part by COSTIND, to form a MEMS state-leading group which will focus fiscal and manpower resources to achieve technical breakthroughs.[19]

Information Acquisition.

China's defense industrial complex is developing a wide range of space- and ground-based sensors to cue long-range precision strike assets and provide advanced warning of impending attacks on its territory. China is slowly progressing toward a redundant, layered, and integrated reconnaissance network. There is a large body of NDU, AMS, and defense industrial writings which strongly advocate that China develop a complete reconnaissance network. A broad survey of PLA and defense industry literature indicates consistent support for the following systems:[20]

- electronic reconnaissance satellites;
- electro-optical reconnaissance satellites;
- synthetic aperture radar satellites;
- missile early warning satellites;
- navigational satellites;
- weather satellites;
- strategic and tactical unmanned aerial vehicles (UAVs);
- airborne early warning;
- space surveillance;
- counterstealth radars;
- SIGINT sites;
- tactical reconnaissance vehicles and ships; and,
- special forces (infiltrated into enemy territory).

Of most significance is that the above list is not just wishful thinking—China has already deployed, or is actively developing every one of these information sources. The defense industries, primarily the electronics and space industry, are developing a broad array of ground-based, airborne, and space-based sensors which Beijing hopes will guarantee battlespace information dominance in any future conflicts around China's periphery.

An overview of the PLA's vast intelligence community is useful to fully comprehend the important role information plays in China's strategy.[21] The focal point for strategic and tactical military intelligence, and counterpart to the U.S. Defense Intelligence Agency, is the GSD's Second Department (*zongcan erbu*). The Second Department oversees military human intelligence (HUMINT) collection, widely exploits open source materials, fuses HUMINT, signals intelligence (SIGINT), and imagery intelligence data, and disseminates finished intelligence products to the CMC and other consumers. Preliminary fusion is carried out by the Second Department's Analysis Bureau which mans the National Watch Center, the focal point for national-level indications and warning. In-depth analysis is carried out by regional bureaus.[22]

Signals Intelligence (SIGINT). Closely linked to the Second Department is the GSD Third Department and the GSD Electronic Countermeasures (ECM) and Radar Department (*dianzi duikang yu leida bu*), China's answer to the U.S. National Security Agency. China maintains the most extensive SIGINT network of all the countries in the Asia-Pacific region. SIGINT systems include several dozen ground stations, half a dozen ships, truck-mounted systems, and airborne systems. Third Department headquarters is located in the vicinity of the GSD First Department (Operations Department), AMS, and NDU complex in the hills northwest of the Summer Palace. The Third Department (*zongcan sanbu*) is allegedly manned by approximately 20,000 personnel, with most of their linguists trained at the Luoyang Institute of Foreign Languages.[23]

SIGINT sites for the collection of radio and SATCOM are spread throughout China with the net control station situated in Beijing. Beijing also has several other SIGINT stations as well. At one time, a site in Lanzhou was responsible for monitoring Russian signal traffic and for providing strategic early warning of a Russian missile attack. The Shenyang station covers signals from Russia, Japan, and Korea. The Chengdu SIGINT site controls the Third Department's operations against India, Pakistan, and Southeast Asia. The Nanjing site monitors Taiwan signal traffic, and the Guangzhou site covers Southeast Asia and the South China Sea. Other sites are located near the Sino-Russian and

Sino-Mongolian border at Jilemutu, Erlian, and Hami. Several sites are in Northwest China. There are at least two SIGINT stations in Shanghai.

Outside China, a SIGINT station has been established on Rocky Island (*Shidao*), near Woody Island in the Paracels. There have been persistent press reports of Chinese electronic surveillance sites in Burma, an ideal location for monitoring naval traffic in the Indian Ocean. China has also established multiple SIGINT sites in Burma and Laos. The Third Department and the Navy cooperate on shipborne intelligence collection platforms. Air Force SIGINT collection is managed by the PLAAF Sixth Research Institute (*kongliusuo*) in Beijing.[24]

The GSD ECM and Radar Department (GSD Fourth Department) has the electronic intelligence (ELINT) portfolio within the PLA's SIGINT apparatus. This department is responsible for electronic countermeasures, requiring them to collect and maintain data bases on electronic signals.[25] ELINT receivers are the responsibility of the Southwest Institute of Electronic Equipment (SWIEE). Among the wide range of SWIEE ELINT products is a new KZ900 airborne ELINT pod. The GSD 54th Research Institute supports the ECM Department in development of digital ELINT signal processors to analyze parameters of radar pulses.[26]

To augment its ground-based collection, China may be resurrecting an ELINT satellite program which has been dormant for over 20 years. The PLA experimented with ELINT satellites, euphemistically called "technical experimental satellites" (*jishu shiyan weixing*), in the mid-1970s under the Shanghai Bureau of Astronautics' 701 Program. The first Chinese ELINT satellite was launched from Jiuquan in July 1975 on the FB-1 launch vehicle which was specifically designed to meet the weight and orbital accuracy requirements of ELINT platforms. The FB-1 launched two more experimental ELINT satellites in December 1975 and August 1976. For unknown reasons, the program was discontinued. Technical writings, however, provide strong indications that the Shanghai Academy of Spaceflight Technology (SAST), the successor of the Shanghai

Bureau of Astronautics, has resurrected the program and intends to field a satellite-borne electronic reconnaissance system. The SWIEE may be developing the ELINT receivers. At least one SAST design under evaluation is a constellation of small electronic reconnaissance satellites which can ensure precise location data and survivability.[27]

Photoreconnaissance. Another source of intelligence information which is increasing in importance is space-based photoreconnaissance, euphemistically referred to as remote sensing (*yaogan*). China is receiving regular satellite intelligence data, building up their digital imagery databases, and gaining experience in the management and interpretation of satellite imagery data. Imagery for use in China's intelligence community comes from two sources: indigenously developed space photoreconnaissance platforms and imagery acquired from foreign sources.

China has been slow to develop a space-based reconnaissance capability. Its first experimental imagery system was launched in November 1975, and was followed by two more tests. The system reached IOC in September 1987 when the FSW-1 (*fanhuishi weixing*, or recoverable satellite) was launched from Jiuquan Space Launch Center, and returned to earth with its film in Sichuan. The FSW-1 provided for wide area imaging and orbits for 8 days. Four FSW-1 were successfully launched between 1987 and 1992. In 1993, a problem in its attitude control system resulted in a failed FSW-1 mission.[28]

The follow-on FSW-2 satellite carries 2000 meters of film and has a resolution capability of at least 10 meters. The first FSW-2, also known as the Jianbing-1B, was launched in August 1992, with subsequent launches in 1994 and 1996. One of the more significant aspects of the FSW-2 is its demonstrated maneuvering capability. The FSW-2 orbits for 15 or 16 days before returning to earth with its imagery package.[29] On October 20, 1996, using the LM-2D from Jiuquan, China launched another "scientific survey" (*kexue shence*) satellite which orbited 15 days before returning to earth.[30] The 1996 FSW-2 launch was expected to be the last in this series as China moves to a more advanced imaging system.[31]

China's remote sensing community is working feverishly to deploy at least three different electro-optical remote sensing platforms. First, China's third generation of imaging satellite, the FSW-3, is expected to be a recoverable system with a one meter resolution. China's Academy of Space Technology (CAST) engineers have also conducted design work on a tactical imagery system and associated mobile ground receiving stations. The system is based on small satellite technology, uses a charged coupled device (CCD) array, and, when operating in a 700-kilometer sun synchronous orbit, is designed to have a five meter resolution.[32] Another series of reconnaissance satellites, jointly developed with Brazil, is known as Ziyuan-1, or ZY-1. The ZY-1 will have a 2-year lifespan and will incorporate a data transmission system to beam images back to earth. The ZY-1, operating at an altitude of 778 kilometers, is expected to have only a 20 meter resolution, but will add to China's experience base in electro-optical imaging systems.[33]

The National Remote Sensing Center, directed by Zheng Lizhong, oversees China's remote sensing community, coordinates requirements, and manages the procurement of imagery from foreign sources. The GSD Second Department's Technology Bureau serves as a PLA representative within China's remote sensing community.[34] The Center's Remote Sensing Ground Station is currently China's sole imagery downlink site, serving over 500 organizations throughout the country. The ground station provides 24-hour a day, year-round service and has a reception footprint encompassing 85 percent of China, all of Korea, Taiwan, and Japan.

The National Remote Sensing Center and its ground station have taken advantage of the relaxation of controls on marketing of high resolution imagery and have contracted with numerous foreign entities to receive remote sensing data. China has received data from U.S. LANDSAT and Russian remote sensing platforms for several years.[35] The ground center has contracted with France's National Center for Space Studies (CNES) to receive SPOT imagery up to a resolution of about 2.5 meters and wide image field of 100 kilometers.[36] China is negotiating with or has contracted to receive data from Canada's RADARSAT, the U.S. LANDSAT 7, and Israel's EROS-A one meter resolution system.[37] The station will also receive downlinked images from the first China-Brazil Earth Resources Satellite (CBERS), slated for launch in 1998. Recent reporting also indicates China may be negotiating with Russia to upgrade its remote sensing relationship to include transmission of real-time high resolution imagery for military purposes.[38]

The ground station, working in conjunction with COSTIND, is rapidly progressing in its capacity and technical capabilities. During the Spring of 1992, Italy's Telespazio signed an agreement with China's National Remote Sensing Center worth $8 million to provide Olivetti image processing computers and software. In March 1992, Telespazio assigned technicians to train Chinese photo interpreters for up to 3 years.[39] In the coming years, the station will be working on ultra-high-speed data processing, storage, and transmission systems, as well as computer, data compression, and networking technology to be able to handle real-time, high resolution imagery from multiple satellites. COSTIND is installing a real-time image storage system as well as an imagery dissemination system, linked to China's national integrated telecommunications network, that will allow subscribers to search and rapidly download images.[40] In 1996, COSTIND installed a digitized high resolution imagery processing system, the BGC-161.[41]

The Institute of Remote Sensing Application (IRSA) is the primary R&D arm of the National Remote Sensing Center. The decision to establish the IRSA was made at the 1978 National S&T Conference, and doors were opened in 1980. IRSA employs almost 300 people and has five basic research departments, three technology research departments, and two applications departments. IRSA manages the Center for Airborne Remote Sensing and the Computer Applications Center. A wide range of recent upgrades to China's remote sensing capabilities are under Project 724. Under the 863 Program, IRSA also has worked to establish parameters for

China's first indigenous SAR satellite. Testing on subsystems has already been completed.[42]

China is moving toward augmenting its electro-optical sensors with synthetic aperture radar satellite systems for all weather, day and night coverage of ground and maritime targets. The PLA and other parts of the state apparatus view radar satellite imagery as critical in China's ability to achieve information dominance. Unlike electro-optical systems, radar satellites, according to GSD Second Department advocates, can see through clouds, rain, and fog in order to detect targets on the ground or underground, and in or under the ocean. In addition, SAR satellites are extremely useful in tracking moving targets, and can be useful in satisfying military mapping requirements.[43] Chinese engineers have been examining SAR satellites as a means to track enemy submarines in shallow waters.[44] China has already fielded a real-time airborne synthetic aperture radar (SAR) system, and is working toward deployment of an indigenous space based SAR satellite (*hecheng kongjing leida weixing*).[45]

Until an indigenous system can be launched, however, China has arranged to receive downlinked radar satellite imagery to help establish a foundation for radar satellite imagery exploitation. China's Remote Sensing Ground Station director Fan Xizhe has contracted with RADARSAT International and the Canadian Space Agency to receive RADARSAT imagery. The Canadian Space Agency has been working with China on various RADARSAT and SAR programs since 1993, under the GlobeSAR project. Furthermore, COSTIND and CASC are negotiating with Canada's Spar Aerospace to construct two RADARSATs. According to the July 1996 draft contract, the first will be launched within the next 3 years and the second by 2004. The satellites will incorporate synthetic aperture radars which can provide all weather, day and night ocean surveillance imagery data.[46]

Canadian RADARSATs have provided up to a 7-10 meter resolution in all weather conditions, operating at an altitude of 800 kilometers. In the northern latitudes, the satellite can image targets every 3 days. Although its primary use is likely civilian in nature, the RADARSAT could allow China to locate ships at sea or large battle groups operating in the Pacific and Indian Oceans, detect military construction, and key other assets for a closer look at militarily significant targets.[47]

China's procurement of a synthetic aperture radar satellite, to include the processing of its images, will lay the foundation for development of an indigenous space based SAR system. Preliminary R&D (*yuxian yanjiu*) on a space-based SAR satellite was initiated in the late 1980s, and model R&D (*xinghao yanzhi*) began in 1991. In May 1995, SSTC and COSTIND approved the finalized design and work on associated high speed data transmission is well underway. In accordance with the national defense S&T strategy, while the first generation SAR satellite is in the prototype development phase, preliminary research has already begun on the second generation SAR satellite system. Key institutes involved in the indigenous development of synthetic aperture radar satellites include CAS' Institute of Electronics, CAST's 501st and 504th Research Institutes (Xian Institute of Space Radio Technology), Shanghai Institute of Satellite Engineering, and MEI's 14th Research Institute and the Southwest Institute of Electronic Equipment (SWIEE).[48]

To gain a real or near real-time indigenous photo-reconnaissance capability, China will need to take one of two potential routes. The primary difficulty is that remote sensing satellites must be within line of sight to download imagery data. Essential for the efficient processing of downlinked imagery is data compression technology, which CAST is attempting to master.[49] The first route which China is working toward is the development of data relay satellites (*shuju zhongji weixing*). Included in China's long-range plan for space development, data relay satellites will allow China's space sensing platforms to pass data to a ground station without being within line of sight of a ground station.[50] The second option is establish ground stations abroad. China signed agreements with France (1993) and Chile (1994) for joint use of their ground stations.[51] Seeking to transmit imagery directly to theater and field commanders, China's remote sensing community has also begun to explore development of mobile remote sensing ground stations.[52]

Space Surveillance. China has an extensive space and missile tracking network. The construction of the network, known as the 154 Project, took place in the 1960s and supported China's missile testing. Subsequent upgrades, undertaken to support the launch and control of the first Dongfanghong (DFH) communications satellite, were done under the 157 Project. Leading figures in this effort included Chen Fangyun, Shen Rongjun, and Wang Daheng. The network has optical, radar, telemetry, and communications components. Software development and systems integration work is done at COSTIND's Luoyang Institute of Tracking, Telemetry, and Telecommunications (LITTT).

COSTIND operates the national space and missile tracking command center in Beijing. The satellite tracking and control network is managed from the Xian Satellite Control Center in the eastern edge of the city. The national satellite control center was formerly located in Weinan, Shaanxi province, but moved to Xian in December 1987. Tracking stations supporting the national network are located at Weinan (GEO/LEO), Xiamen (GEO/LEO), Nanning (LEO); Kashgar (LEO); Changchun (LEO); and Minxi (GEO/LEO). China's Academy of Sciences Nanjing Observatory does orbital predictions and modeling. China also operates shipborne missile tracking platforms (*Yuanwang*), which are capable of operating throughout the Pacific, Atlantic, and Indian Oceans. In addition, China is working to link up with foreign tracking and control networks, and has signed agreements for space tracking cooperation with Chile, Kiribati, and France. Chinese engineers are also working to establish a space-based component to their TTC network.[53] The current network is primarily to support its own satellites and has only limited capabilities against non-cooperative spacecraft. However, it serves as the foundation for future efforts to develop a more robust space tracking system which can accurately track foreign systems as well. Beijing may have already received technical assistance in laser radars used to track image satellites and may be seeking an advanced radar system with the ability to track satellites in low earth orbit.[54]

Over-the-Horizon (OTH) Sensors. China began development of HF ground wave OTH sensors in November 1967. A group led by Deputy Director of the Space and Missile Industry Qian Xuesen was assigned the responsibility of developing a ground wave OTH sensor able to detect targets at 250 kilometers. The radar was to provide targeting data for their embryonic anti-ship cruise missile program. In the 1970s, an experimental ground wave OTH radar, with an antenna length of 2300 meters, was deployed. Foreign export restrictions, however, prevented China from obtaining technology which was needed for further improvements.

Since 1985, developments in computing, microelectronics, and digital signal processing have permitted further advances in China's OTH technology. Chinese institutes, such as the Harbin Institute of Technology (HIT), have most recently concentrated on improving the OTH sensor's ECCM capabilities and digitizing the radar system. There is also some interest in developing a mobile version of the OTH radar. China has also developed an experimental sky-wave OTH radar which tracks aircraft targets at a range of 1,000 kilometers. Chinese systems engineers believe the OTH system can contribute to an integrated air defense system. In 1995, HIT tested a new HF radar system capable of detecting low altitude and sea-skimming targets as well as naval targets.[55]

Phased Array Radars. Chinese engineers see increased opportunities for indigenous development of solid state phased array radars due to advances in integrated circuit technology. The MEI's 14th Institute, located in Nanjing, is China's leading organization for phased array radar development. An integral part of China's space tracking network includes large phased array radars (LPAR). Work on LPARs began in 1970. They were intended to catalogue space targets and provide for early warning of missile attacks. At least one radar, positioned on a mountain slope at 1,600 meter elevation near Xuanhua, is believed to be manned by Second Artillery forces.[56] The 7010 radar was developed by MEI'S

14th Research Institute. The antenna array was completed in 1976 and has gained prominence for its tracking of several satellites. The 7010 radar hands off targets to other tracking sensors, such as the 110 radar. The 110 radar is a large monopulse precision tracking surveillance radar which began R&D in 1966. The radar system, enclosed in a 44 meter-in-diameter radome, was commissioned into service in 1977. Like the 7010 radar, the 110 radar was developed by the 14th Research Institute in Nanjing.[57]

Unmanned Aerial Vehicles (UAV's). Chinese military officials have also given priority to the development of UAV's (*wuren jiashi feiji*). UAV R&D is centered in BUAA's Institute of Unmanned Flight Vehicle Design Institute and NUAA's Institute of Unmanned Aircraft. One of China's newly deployed UAV's is the ASN-206, developed by the Xian ASN Technology Group. This UAV has a coverage range of 150 kilometers, can operate both day and night, and can carry a variety of sensors, including optical and infrared. Real-time intelligence can be transmitted to a ground control station for targeting purposes. Maximum altitude of the system is 5,000-6,000 meters, and its flying time is 4-8 hours.[58] NUAA is designing a GPS-based navigation system to assist in route planning and course navigation and a new UAV which incorporates a low radar cross-section design.[59] Beijing may also have concluded an agreement with a foreign supplier to acquire a high altitude, long endurance UAV, together with a ground control station, and either production or co-production rights. This UAV would give China the ability to conduct extended imagery reconnaissance and surveillance, ELINT collection, as well as electronic warfare missions.[60]

Development of China's National Telecommunications Infrastructure.

The vast amount of data collected by China's intelligence system requires a highly advanced national telecommunications infrastructure. The Chinese leadership clearly recognizes that a modern, survivable C^4I network is an essential element in modern and future warfare. A high capacity communications network, capable of handling multiple gigabits per second, is vital in linking targeting data with strike assets. Over the last 5 years, China has opened its market to the leading telecommunications firms around the world in what is truly an information revolution. As AT&T official Bill Warwick stated, "China has the opportunity to leapfrog almost overnight into the information age."[61] The emphasis is on high capacity fiber optics; switching systems, to include highly survivable common channel signaling No. 7 software; satellite communication systems; and systems integration and data fusion.

China has taken initial steps toward forming a national integrated C^4I system. Before 1991, China's C^4I system was disjointed and void of interconnectivity. After the Gulf War, the State Council and CMC directed the establishment of a national information system which will bring together various communications networks. The Eighth Five-Year Plan (1991-95) marked a dramatic change in China's attitude toward telecommunications, which has become the fastest growing sector of China's dynamic economy. The State Council directed the formation of an Information Technology Leading Group which develops policy and coordinates information technology development. By the end of 1995, China had constructed ten of the largest networks in the world. Based on Beijing's long range telecommunications plan to 2050, this rapid modernization is expected to continue. This full court press toward informationalization has prompted some to assert that China has the potential to become the world's most advanced telecommunications infrastructure.[62]

Foreign providers have been key to China's initial successes in telecommunications development. Leading companies include Alcatel, Ericsson, Siemens, Nokia, Northern Telecom, AT&T, Sprint, and Motorola. Providers of communications satellites and associated technology include Hughes (primarily very small aperture terminals, or VSATs), Lockheed-Martin, and Loral.

The military implications of China's revolutionary leap in information capabilities are significant. Recognizing the need for a real-time sensor-to-shooter capability, the PLA is working closely with the Ministry of Post and Telecommunications (MPT), the electronics industry, and the space industry to help establish an integrated high capacity national information infrastructure. Since promulgation of the 863 Program, China's telecommunications community has embarked upon a long-term, dual-use program valued at $200 billion U.S.[63]

China's telecommunications sector encompasses civilian and military organizations. As is the case in the United States, a national telecommunications network is inherently dual-use, meaning civilian and military sectors can benefit from its use. The MPT, under the leadership of the State Council, is the preeminent player in China's telecommunications sector. MPT held a monopoly over telecommunication services until 1994, when the State Council authorized the formation of a competing network, United Telecommunications Corporation (Liantong). Liantong came into existence through the merging of the electronics, railways, and power industries, and at least a portion of the PLA General Staff Department communication networks.[64] MEI's China Academy of Electronic Information Technology also plays a role in developing a national information infrastructure.[65]

Development of dedicated PLA communications networks is a top national priority. Since 1992, the capacity of PLA communication networks has increased 10-fold. According to one estimate, the central government allocates approximately 20 percent of the total telecommunications budget toward dedicated PLA communications systems.[66] There are currently at least four military networks—a military telephone network, a secure telephone network, an automated command system, and an integrated field communication network. The automated command system links ground force units with PLA Air Force and Navy units. Construction began in 1987 and, when completed, will be able to transmit data, imagery, and other types of militarily useful information.[67] PLA and civilian networks are currently separate. There are efforts underway, however, to integrate PLA telecommunications networks into the public communications network, making extensive use of civilian services possible.[68]

The PLA is developing its communications infrastructure in accordance with the "six transformations." These include moving from analog to digital technology; electrical cable to optical cable; electromechanical switches to program-controlled switches; single function terminals to multi-function terminals; single service networks to networks offering integrated services; and administration based on human labor to control based on automation and the application of intelligent technology.[69]

Two of the leading PLA organizations responsible for development of China's defense information infrastructure include the GSD's Telecommunications Department and COSTIND's Beijing Institute of Systems Engineering (BISE). Entities under the Telecommunications Department include the GSD Information Engineering Academy and a communications-related business endeavor, the China Electronics Systems Engineering Company (CESEC), which operates what is commonly known as China's third communications network.[70] BISE works closely with the GSD Telecommunications Department and civilian entities in integrating various networks into an national system.[71] PLA communication troops also work with national and provincial MPT authorities, often being tasked to install communications links.[72] China's investment into a modern information infrastructure is nothing short of revolutionary. Purchasing some of the world's most advanced telecommunications technology, China is installing a dizzying array of terrestrial and satellite networks.

China's telecommunications development will provide a significant boost in the PLA leadership's ability to command and control its forces. The central headquarters of the PLA GSD C^4I apparatus is located in the Xishan area in the western suburbs of Beijing. It functions as a communications, intelligence, and combat control center. In short, the Xishan command complex is the operational nerve center of the PLA ground units, air forces, navies, and strategic missile forces, similar in nature to the Pentagon's National Military

Command Center. The Command and Control Headquarters includes the GSD First Department (known as *zongcan zuozhan bu*, or operations department), key Second Department offices, and the Third Department. The First Department, under direct control of the GSD director, mans the Command and Control Headquarters 24 hours a day.

All operational orders to PLA units originate from the Command and Control Headquarters. Mobilization orders usually result from high level meetings held in the Xishan complex, attended by the CMC Standing Committee, other high ranking military officials, and selected civilian leaders. Orders can be transmitted through a number of means, including secure telephone or landline. The military command network is a closed network. Orders from the national command authority can go to the military region or directly to an individual division.[73] There is an effort underway to develop a strategic SATCOM network which uses very small aperture terminal (VSAT) mobile ground stations with an antenna smaller than three meters in diameter. Eventually, all units at the group army-level and above will be equipped with this capability.[74]

The Ground Segment. The mainstay of China's communications development is installation of a backbone fiber-optic network, linking all provincial capitals except Lhasa. With work well underway to connect cities at the prefecture level, installation of fiber-optics has achieved an annual growth rate of 92 percent. The fiber optic network augments previously existing microwave and high frequency (HF) networks.[75] To ensure various networks can accommodate vast amount of data, MPT, ostensibly backed by the PLA, is investing heavily into synchronized digital hierarchy (SDH), asynchronous transfer mode (ATM), synchronous optical networking (SONET), wavelength division multiplexing (WDM), and other technologies which significantly increase survivability and useable bandwidth.[76]

China's national fiber-optic network is an essential aspect of China's quest for information dominance. Fiber-optics deny an adversary the ability to intercept communications since there is no transmission of signals through the air. The only viable means of tapping into fiber-optic links is to physically penetrate the cable, which can be detected by system administrators. Fiber-optic networks also complicate the ability of overhead photo reconnaissance platforms to detect command and control nodes, which usually have a noticeable signature. These types of technologies associated with fiber-optics are capable of "all optical" operations in that there are no optical-to-electronic transitions which may cause bottlenecks or limit versatility. These networks are also rapidly reconfigurable in the case of a sudden surge in communications requirements.[77]

The Space Segment. COSTIND, MPT, and China's space community are making progress in integrating a space segment into its overall telecommunications infrastructure. All four of their first operational generation satellites, the DFH-2 series launched in the 1980s, have ceased to function. After a long delay, CASC and COSTIND have succeeded in placing the first of their second generation domestic communications satellite, the DFH-3, in orbit. The first DFH-3, launched in November 1994, failed to reach its proper orbit. However, the second attempt to place a DFH-3 into orbit in May 1997 was successful.[78]

CASC's satellite R&D and manufacturing arm, CAST, has commenced work on the next generation communication satellite, the DFH-4, a direct broadcasting system which can transmit data to users without the need for ground station rebroadcasting.[79] The fielding of a direct broadcast satellite offers a capability for distributing information to the lowest echelon in a battlefield. The term "sensor-to-shooter" can become a reality if this technology were adapted to military applications, potentially transmitting data (maps, pictures, and enemy deployments) on demand to small units, each using an 18-inch (or smaller) capability to receive orders and situational information. The DFH-4 could feasibly permit the transmission of tailored data to hundreds of units simultaneously. CAST is also researching measures to ensure survivability of their satellite communications, to include spread spectrum and frequency hopping technology.[80]

While developing an indigenous ability to build communications satellites, COSTIND and MPT have contracted to purchase several foreign satellite systems. To fill the gap in coverage after the DFH-2 satellites approached the end of their service life, MPT purchased an in-orbit satellite, CHINASAT-5, in 1993. MPT, COSTIND, and CASC also joined in a partnership arrangement with Thai and Hong Kong companies to form Asia-Pacific Telecommunications Company (APT). Their first satellite, APSTAR-1, was launched in July 1994 and is expected to have a lifetime of 10 years. APSTAR-2 exploded shortly after launch in January 1995. A replacement satellite was launched in April 1997.[81]

Other satellite procurements included CHINASAT-7 which was launched in August 1996, but failed to reach its proper orbit. China has contracted for a replacement satellite with 52 transponders (36 C-Band/16 Ku-Band), the CHINASAT-8 (Zhongwei-8), which will be launched by the end of 1998.[82] In August 1995, China Orient Telecommunications Satellite Company signed a contract for a U.S. manufactured satellite, designated as the CHINASTAR-1 (Zhongxing-1), which will be launched on an LM-3B during 1998. This modern, high capacity system of 38 transponders will provide voice, data, and other services, and will have a lifespan of 15 years. China Orient's TT&C ground station will be located within a complex of existing ground stations in Dongbeiwang, northwest of the Beijing's Summer Palace.[83] COSTIND, CASC, and MPT are also working closely with Germany in the joint development of another generation of communication satellite, SINOSAT-1 (Xinnuo Weixing-1), the first of which is due for launch in 1998. The German firm Teledix is serving as a subcontractor for the SINOSAT-1, the DFH-3, and other programs, supplying attitude control systems to orient and stabilize the spacecraft body.[84]

China is also participating in international consortiums which will provide global mobile telephone services. First, CASC is a shareholder in the Iridium consortium, which will provide global cellular services via a 66 satellite constellation. COSTIND will launch 22 of the satellites. An Iridium gateway in China will interconnect with the public switched network which will make communications possible between Iridium lines and any other telephone in the world.[85] COSTIND has also given serious consideration to developing its own global network, the Global Mobile Satellite Information System (GMSIS). GMSIS, utilizing a small number of satellites in medium earth orbit, will provide mobile communications services to a large number of domestic and regional customers.[86]

Finally, COSTIND, CASC, Liantong, and MPT have signed an agreement with Singapore and Thailand for joint procurement and management of a communications satellite which will offer mobile hand-held phone service. The consortium, called Asia-Pacific Mobile Telecommunications (APMT) will buy a U.S. satellite for U.S. $640 million. Launch of an LM-3B from Xichang is scheduled for mid-1998. The initial concept for the system called for a Hughes 601 platform in geostationary orbit off the west coast of Sumatra. The system, which will operate in the L and Ku bands, is designed to handle 4,000 calls simultaneously. Chinese requirements include multiple beam technology (specifically 32) for increased security.[87] China's space and ground communications segments are being integrated into a national information infrastructure.[88]

Systems Integration. Combining the various networks and equipment into a survivable, integrated network is a national priority. Concerned about potential effects of an enemy strike against a communications node, China's telecommunications community is investing heavily into common channel signaling number seven switching technology (SS7). SS7 software allows for automatic rerouting of traffic should a key C3 node be disabled. Beijing hopes to complete a national SS7 "intelligent network," within the current 5-year plan (1996-2000). MPT and the PLA Telecommunications Department's Institute of Information Engineering jointly developed China's first domestically produced SS7 intelligent network, the CIN-01, and in March 1997, the PLA GSD installed an intelligent SS7 network in the Beijing area.[89] As a side note, much the PLA's software development for systems integration is concentrated within COSTIND's Beijing Institute of Systems Engineering.[90]

On the military side, one concept called for establishing and connecting five military networks—command and control systems, intelligence and reconnaissance, early warning and detection, communications, and electronic warfare. In theory, the formation of a defense information infrastructure will allow operational integration and combined command of China's ground, naval, air, and Second Artillery forces. Sources indicate the PLA currently does not have a comprehensive C^4I network, but appears to be adopting a step-by-step approach toward forming one.[91]

Computing. Dominating the information environment requires the computer processing of vast amounts of data. High speed computers are also essential for simulations and modeling, and operation of a wide range of new weapons systems. Recognizing the importance of computing, the State Council formed a state leading group on computerization in May 1996. The leading group is headed by Zou Jiahua and consists of representatives from MEI, MPT, State Planning Commission, People's Bank of China, and 12 other departments. The group will formulate national policies, strategies, and mid- to long-range plans for national computerization.[92]

COSTIND and other state R&D organizations have prioritized development of supercomputing technologies under the 863 Program. The SHUGUANG 1000 parallel computer system, an 863 funded program, achieved 2.5 billion operations per second in 1985. Projects underway include a CAS effort to develop a parallel supercomputer system, the DAWN series, capable of speeds up to 300 billion calculations per second.[93] China's defense industry is also working to reduce the size of its microcomputers for missile, launch vehicle, and satellite use. For example, CASC's first microcomputer for launch vehicles weighed 50 kilograms. The size was reduced to 30 kilograms, and reduced again to 10 kilograms.[94]

China is increasing emphasis on computer simulation. Useful systems include a new series of Chinese supercomputers, the Yinhe (Galaxy), developed by COSTIND's National University of Defense Technology (NUDT) in Changsha, Hunan province. Most recently, NUDT developed an advanced real-time simulation system which has been used by CASC and the aviation industry in the development of a new generation of missile control systems and UAVs.[95] CASC's primary simulation facility, the Beijing Simulation Center, is the largest in Asia. As a side note, in April 1997, the Russian Defense Minister initiated cooperative efforts with the Beijing Simulation Center in computer simulation technology.[96]

China is also investing heavily in the development of artificial intelligence and intelligent computer systems. The SSTC and COSTIND designated intelligent computing as a key area of development under the 863 Program. National level coordination of artificial intelligence R&D is done through the Expert Group for Intelligent Computing (code named 863-306), chaired by Wang Chengwei. Establishment of the National Research Center for Intelligent Computing Systems further reflects growing emphasis on artificial intelligence.[97] Artificial intelligence systems, such as neural networks, have a wide variety of applications, to include weapon system design and manufacturing, simulation, automatic target recognition (ATR), and decisionmaking.

China is poised to become a world leader in advanced computer networking technology. According to AT&T Chairman Louis Gerstner, "China has the opportunity to move more quickly than other countries because its has invested less in traditional systems and practices." China's across-the-board introduction of state-of-the-art systems avoids pitfalls other countries are facing, such as investing in making incompatible computer systems talk with each other.[98]

Information Attack.

A final aspect of China's quest for information dominance is information attack. PLA strategists see precision targeting of critical command and control nodes, computer warfare, and a counterspace capability as means to offset adversarial strengths. The development of an effective electronics countermeasures capability is a top priority within the PLA General Staff Department, specifically the Electronic Countermeasures and Radar Department (*dianzi duikang yu leida bu*), and within the R&D community. Information warfare advocates are pressing for more sophisticated means to engage in command and control warfare, including better jammers and high powered microwave technology.

First, PLA strategists are increasing emphasis on computer warfare as one means to offset adversarial strengths. Chinese observers point out the effectiveness of U.S. use of viruses against Iraqi computer systems during the Gulf War and note the utility of using computer viruses as a weapon. Engineers have conducted feasibility studies on introducing viruses (*bingdu*) into adversary's computer systems from long distances via wireless means. Besides serving as a means to attack computer networks, these studies have been useful in developing countermeasures to enemy attempts to inject viruses into Chinese computers.[99]

China's defense industrial complex is placing a great deal of emphasis on electronic countermeasures (ECM) development. The GSD Electronic Countermeasure and Radar Department (also known as GSD Fourth Department), established in 1990, has overall responsibility for electronic warfare, including electronic intelligence collection and maintenance of threat libraries and electronic orders of battle. Besides coordinating the PLA electronic warfare doctrine and strategy, GSD Fourth Department units provide electronic warfare defense of strategic targets, such as the PLA command bunkers in the Western Hills (*Xishan*) of Beijing.[100] The GSD Fourth Department leadership is currently tackling problems associated with command and control of ECM operations. Because of the close coordination required, an

ECM command operations cell will serve as an integral part of a theater command post.[101] Electronic warfare units have conducted major joint exercises as recently as March 1996 in the South China Sea.[102]

The research institute most responsible for ELINT development and radar jammers is the electronic industry's SWIEE (the 29th Research Institute) in Chengdu, Sichuan province. Research and development on communications jammers is the responsibility of the 36th Research Institute in Hefei, Anhui province. SWIEE is also conducting operational feasibility studies on using UAVs as electronic warfare platforms.[103] The PLA's primary academic and training institute on electronic warfare appears to be the PLA Academy of Electronic Engineering, located in Hefei, Anhui province.[104]

Chinese planners have begun to develop electronic warfare concepts to counter adversarial air and space systems as well. For example, there are indications the PLA is preparing to test or deploy a SATCOM jammer.[105] SWIEE engineers have carried out feasibility studies for jamming synthetic aperture radar reconnaissance satellites.[106] Chinese research institutes are also conducting feasibility studies on jamming the U.S. NAVSTAR Global Positioning System.[107] Finally, Chinese engineers are examining methodologies to enable the PLAAF to jam airborne early warning platforms and sophisticated networks, such as the Joint Tactical Information Distribution System (JTIDS).[108]

One of the most significant foreign sources of electronic warfare technology is Israel. In the late 1980s/early 1990s, Israel export firms and China's Ministry of Electronics Industry signed a memorandum of understanding on technological transfer of electronic warfare hardware and software. Israeli companies such as Elbit, Elisra, Tadiran, Elop, and Elta entered into cooperative development ventures with Chinese entities such as SWIEE, the Hebei-based Communications, Telemetry, and Telecontrol Research Institute (MEI 54th Research Institute), and the Anhui-based East China Research Institute of Electronic Engineering (ECRIEE, or the MEI 36th Research Institute). Projects include airborne radar warning receivers, ELINT systems, and radar jammers.[109]

Under COSTIND guidance, China's R&D community is also dabbling in even more advanced information attack technology related to high powered microwave (HPM) weapons. China's most respected advocates of information warfare have strongly supported HPM development. COSTIND subordinated research institutes have already mastered certain power sources commonly associated with microwave weapons, including flux compression generators (*baoci yasuo fashengqi*) and vircators (*xuyin jizhen dangqi*). Chinese writings have outlined three attack applications for HPM devices: as a directional microwave air defense system to damage the electronic systems of attacking aircraft and antiradiation missiles (known to Chinese engineers as an electromagnetic missile); an ASAT weapon to degrade sensitive satellite electronic systems; and a warhead to use in an opening salvo to shut down adversarial radars and C^4I systems.[110]

In addition to "soft kill" approaches to information dominance, China's R&D establishment is developing a variety of "hard kill" means to neutralize segments of an enemy's C^4I infrastructure. Employment of ballistic, cruise, and antiradiation missiles against C^2 centers and ground-based early warning sites is one approach. Targeting of airborne early warning platforms is another. The Chinese defense industrial complex is examining a range of designs for a long-range air-to-air missile to counter high value assets such as airborne early warning platforms, airborne warning and control systems (AWACS), and airborne jammers. Development appears to be in the theoretical evaluation (*lunzheng*) phase where various institutes evaluate system feasibility and performance requirements. Key organizations in this effort include the Shanghai Academy of Space Technology's primary missile design institute (Shanghai Institute of Electromechanical Engineering) and the PLAAF Eighth Research Institute (Weapon System Evaluation). A solid propulsion system and combined passive microwave and infrared imaging seekers are under consideration.[111]

The most common concept for a long-range anti-airborne early warning (AEW) or anti-AWACs system is to use a missile with a jet engine, plus a rocket booster for final approach to the airborne platform. A missile, preferably with a low radar cross section, could approach the AEW platform at an extremely low altitude using passive radar, homing in on the aircraft's radar signal. Once under the AEW platform, it would ignite its rocket and home in on the AEW using active radar or infrared. Such a missile could be quite difficult to avoid, as it might be 20-30 seconds before such a fast moving missile from below connected.[112]

CASC is already marketing a surface-to-air anti-radiation missile system. The FT-2000, intended to neutralize airborne jammers and early warning sensors, operates in the 2-18GHZ range and can reach targets out to an advertised range of 100 kilometers.[113]

A final, less high tech approach to information attack is the growing PLA emphasis on special operations forces. Chinese writings indicate one of the most important of special operations missions is to carry out strikes against an adversary's leadership and command centers.[114] After clandestine entry into enemy territory, special operations troops could strike critical early warning installations, cut communications landlines running out of command centers, or designate targets for air strikes.

Information Protection.

One of the most important pillars in China's quest for information dominance is denying an adversary information on PLA plans, force deployments, and vulnerabilities, and protecting the PLA's ability to command and control its forces. Secrecy has been a traditional aspect of Chinese strategy, and as an authoritarian society, information security is pervasive. However, China's entry into the information age presents a new set of problems. The 1992 failure of 12 railroad computer systems due to viruses was a rude awakening for China's leadership. The rise of information warfare has driven fears that its growing information infrastructure is vulnerable to attack or sabatoge. To help counter its growing vulnerability, Beijing established a National Key Laboratory of Information Security under China's University of Science and Technology.

As a further measure, in 1996, China's State Secrets Bureau and the State Council's Development Research Center published a strategy for information security. The most significant conclusion was a call for the State Council's Leading Group on Information Technology to develop information security strategy and technology in parallel with China national information infrastructure development. The report also points out that China's computer systems, now largely unprotected, need to enhance their security measures. Other sources indicate China has taken steps to protect the leakage of electronic emissions from computer systems which can be intercepted and exploited by adversaries. In 1996, China Academy of Science established the Information Security Technology Engineering Research Center while the State Council began formulation of TEMPEST standards.[115]

Chinese leadership and strategic analysts are becoming increasingly concerned about foreign attempts to insert viruses into their information systems. Acutely aware of the havoc which computer viruses can cause to a command and control system, PLA analysts have advocated measures to protect internal computer networks. These measures include decreasing dependence on foreign sources of integrated circuits, increasing investment into R&D on computer warfare, strengthening information systems' ability to resist EMP, and improving computer security management.[116] At least one PLA institute dedicated to INFOSEC and computer security is the PLA Academy of Electronic Technology in Zhengzhou, Henan province.[117]

In May 1996, *Jiefangjun Bao* (*PLA Daily*), published an article calling for greater attention to emission security. The article was sparked by the April 1996 death of a Chechen separatist leader who was killed when the Russian Air Force plotted his precise location after he used a cellular phone. The PLA has also studied the success of U.S. SIGINT in monitoring and cataloguing Iraqi electronic transmissions before and during the Gulf War. The PLA has issued directives to strictly control any communications and radar emissions and implement an effective frequency management system.[118]

Beijing is examining a wide range of technologies to reduce vulnerabilities of its communications to interception or jamming. For example, widespread introduction of fiber optics communications significantly increases its communications security. Engineers are studying the application of spread spectrum and frequency hopping technology for Beijing's satellite tracking and control network, as well as more secure satellite communications methodologies.[119] China is also stepping up integration of more complex encryption (*mimaxue*) algorithims. At least one organization involved in cryptography is the Sichuan Institute of Information Science, which produces one of China's leading publications on the subject, entitled "Leading Edge Research in Cryptography." A major conference, CHINACRYPT '96, was recently held to discuss the latest advances in encryption.[120]

To augment other information security measures, Chinese engineers and military operators stress concealment, camouflage, and deception (CCD) concepts which are ingrained into traditional Chinese strategic constructs. Writings of industry and military officials indicate widespread attempts to deny foreign satellite platforms information related to disposition of missile forces and other strategic assets. Chinese strategists differentiate camouflage and concealment by the degree of technology involved. Camouflage includes decoys, netting, radar angle reflectors, and other means to prevent an adversary from detecting forces and weapon systems.

Technology, however, has reduced the effectiveness of traditional camouflage measures and has forced China to move toward concealment. Concealment is the employment of technology to decrease detection by adversary satellite systems. Concealment measures, integrated into the weapons R&D process, include electromagnetic, radar, noise, and infrared signature reduction. China emphasizes use of camouflage to avoid detection by foreign optical, infrared, and radar satellites. For example, they camouflage their missile silos and construct dummy silos. Analysts acknowledge that camouflage has become more difficult since the advent of radar and infrared satellites. However, Chinese strategists are certain that emerging concealment materials and technology can effectively counter foreign surveillance systems. One example is infrared suppression, which will counter heat-seeking precision guidance systems.[121] PLA engineers have also published technical papers on methods to reduce infrared signature of underground facilities.[122]

China's efforts to develop counterstealth systems have contributed to the development of low observable technology for aircraft, UAV's, and cruise missiles. China clearly recognizes the advantages of low observable technology and has stepped up investment into stealth (*yinshen*) research, achieving what some engineers believe to be technological breakthroughs.[123]

Psychological operations, manipulation, and strategic deception aspects of information warfare are deeply ingrained in Chinese military doctrine. Traditional Chinese military history, to include the strategic writings of Sun Tzu, is full of examples of strategic and tactical surprise, and deception. Initial Chinese entry into the Korean War can be characterized as a strategic surprise.

In summary, Chinese military planners appear to be striving for a comparative advantage in the ability to control, collect, process, act upon, and disseminate information, giving the PLA a future edge in conflicts around its periphery. The PLA is increasingly viewing information as a strategic weapon. Just as possession of nuclear weapons marked a major power in the 1950s, 1960s, and 1970s, information technology is characteristic of a major power of the 21st century. Crucial in this effort is the development of critical communications and information processing technologies, such as space-based surveillance, direct broadcast satellites, and high speed computers. China's information technology community is making necessary investments into correcting shortcomings in Beijing's ability to integrate complex information systems. At the same time, the PLA is investing in systems which can disrupt an adversary's C^4I infrastructure as a means to blind and disrupt the leadership's ability to control forces under his command.

China's rate of improvement is likely to improve dramatically in the next decade. China's information revolution is driven by technologies available worldwide. Digitization, computer processing, precise global positioning, and systems integration—the technological basis on which a range of new capabilities depends—are available to any country with the money and will to use them systematically to improve military capabilities. Based on its large investment into sensors, telecommunications, electronic warfare, and information protection technologies, China clearly views exploitation of the information revolution as key to its strategic modernization.

ENDNOTES - CHAPTER 3

1. "Guofang Kegongwei Kejiwei Zhaokai Disanjie Nianhui,"(COSTIND S&T Committee Opens Third Annual Meeting), in *Keji Ribao (S&T Daily)*, March 14, 1994. Members also called for prioritizing programs for remote sensing technology, space shuttle, a single-stage-to-orbit space vehicle (SSTO), stealth technology, flexible mirrors (*lingjing*), and materials science. Members present included Zhu Guangya (COSTIND S&T Committee Chairman, physicist with U.S. PhD), Chen Fangyun (tracking and control specialist, chief designer of China's national satellite TT&C network, and COSTIND advisor), Cheng Kaijia (COSTIND advisor, nuclear scientist involved with China's nuclear bomb development), Chen Nengkuan (COSTIND S&T Committee Deputy Director, nuclear physicist with U.S. PhD), and Zhao Renkai (SLBM scientist).

2. Shen Weiguang, "Xinxizhan: Mengxiang Yu Xianshi" ("Information Warfare: Dreams and Reality"), *Zhongguo Guofang Bao (China Defense News)*, February 14, 1997, p. 3; Zhang Feng and Li Bingyan, "Historical Mission of Soldiers Straddling the 21st Century: Roundup of 'Forum for Experts on Meeting the Challenges of the World Military Revolution'," in *Jiefangjun Bao (PLA Daily)*, January 2, 1996, in *FBIS-CHI-96-061*; and Shen Weiguang, *Zhongguo Guofang Keji Xinxi*, September-December 1996, No. 5/6, pp. 87-89, in *FBIS-CHI-98-029*.

3. Shen Weiguang, p. 3; Zhang and Li; and Wang Jianghuai and Lin Dong, "Viewing Our Army's Quality Building From the Perspective of

What Information Warfare Demands," *Jiefangjun Bao*, March 3, 1998, p. 6, in *FBIS-CHI*-98-072.

4. Jen Hui-wen.

5. Cai Renzhao, "Exploring Ways to Defeat the Enemy Through Information," in *Jiefangjun Bao*, March 19, 1996, p. 6, in *FBIS-CHI*-96-100.

6. See, for example, Dai Kouhu, "Accepting the Challenge: Reinforcing China's Defense Information Modernization," *Zhongguo Dianzi Bao*, October 24, 1997, p. 8, in *FBIS-CHI*-98-012.

7. Liang Zhenzing, "New Military Revolution and Information Warfare," p. 8. Over the last few years, Liang Zhenzing has emerged as one of COSTIND's most authoritative figures on C^4I and information warfare issues.

8. Wang Jianghuai and Lin Dong, p. 6.

9. Zhang Deyong, Zhang Minghua, and Xu Kejian, "Information Attack," *Jiefangjun Bao*, March 24, 1998, p. 6, in *FBIS-CHI*-98-104.

10. Xu Sheng and Gao Jianlin, "Shenyang MR Holds Information Warfare Exercise," *Jiefangjun Bao*, March 30, 1998, in *FBIS-CHI*-98-120; for a general PLA overview of information warfare, see Wang Baocun, "A Preliminary Analysis of Information Warfare," *Zhongguo Junshi Kexue*, November 1997, in *FBIS-CHI*-98-093.

11. Most prominent among information warfare specialists is Major General Wang Pufeng, former Vice-President of the Academy of Military Sciences. See Wang Pufeng, pp. 186-189. For an excellent American analysis on Major General Wang's work, see John Arquilla and Soloman Karmel, "Welcome to the Revolution . . . in Chinese Military Affairs," *Defense Analysis*, Vol. 13, No. 3, pp. 255-270; also see Li Qingshan, *Xin Junshi Geming Yu Gaojishu Zhanzheng* (*New Military Revolution and High Tech Warfare*); Chen Huan, "Military Aspects of Information Technology Viewed," in *Xiandai Junshi* (CONMILIT), March 11, 1996, pp. 8-10, in *FBIS-CHI*-96-169; Shen Weiguang, "Focus of Contemporary World Military Revolution—Introduction to Information Warfare," *Jiefangjun Bao*, November 7, 1995, p. 6, in *FBIS-CHI*-95-239; Wang Huyang, "Five Methods of Information Warfare," in *Jiefangjun Bao*, July 16, 1996, p. 6, in *FBIS-CHI*-96-194; Liu Fengcheng and Yu Shuangquan, "Concentrate Forces in New Ways in Modern Warfare," *Jiefangjun Bao*, November 21, 1995, p. 6, in *FBIS-CHI*-96-017; and Hai Lunh and Chang Feng, "Chinese Military Studies Information Warfare," in *Kuang Chiao Ching*

(Wide Angle), January 16, 1996, No. 280, pp. 22-23, in *FBIS-CHI*-96-035. COSTIND's Deputy Director Huai Guomo blessed these basic tenets in a 1996 *China Military Science* article. See Huai Guomo, "On Meeting the Challenge of the New Military Revolution," *Zhongguo Junshi Kexue (China Military Science)*, February 20, 1996, pp. 22-25, in *FBIS-CHI*-96-130; also see Wang Xusheng, Su Jinhai, and Zhang Hong, "Xinxi Geming Yu Guofang Anquan" ("Information Revolution and Defense Security"), *Jisuanji Shijie (China Computerworld)*, No. 30, August 11, 1997, p. 21. Wang, Su, and Zhang are from the PLA Academy of Electronic Technology. Also see Zhang Deyong, Zhang Minghua, and Xu Kejian, p. 6; and Cai Renzhao, p. 6. Also see Lei Zhuomin, "Information Warfare and Training of Skilled Commanders," *Jiefangjun Bao*, December 26, 1995, p. 6, in *FBIS-CHI*-96-036.

12. Ouyang Wen, "The Gulf War Shows the Importance and Urgency of Strengthening Construction of China's Electronic Industry," *Zhongguo Keji Luntan (Forum on Science and Technology in China)*, July 18, 1991, pp. 20-21, in *JPRS-CST*-91-020; Zhang Pingli, "Development of China's Electronics Industry," *Renmin Ribao*, August 5, 1994, p. 1, in *FBIS-CHI*-94-177; and Zhang Xiaojun, "Modern National Defense Needs Modern Signal Troops," *Guofang (National Defense)*, October 15, 1995, pp. 20-21, in *FBIS-CHI*-96-035.

13. AMS, for example, believes reducing the cost of microelectronics is critical. General Mi Zhenyu notes that microelectronics account for 33 percent of the cost for an aircraft, 45 percent in a missile, 66 percent in space vehicles, 22 percent in ships, and 24 percent in ground combat vehicles.

14. Zhang Zhizhong, "Tactical Advantages of Quasi-Wideband Phased Array Radar," *Dianzi Xuebao (Acta Electronica Sinica)*, March 1993, pp. 86-91, in *JPRS-CST*-93-009.

15. At the time of this writing, MEI has been folded into a new Ministry of Information Technology.

16. Hu Qili is a former member of the Politburo Standing Committee. As a close associate of Zhao Ziyang, he was removed from his Standing Committee position in 1989. He was rehabilitated in 1993 and appointed to reinvigorate the electronics industry.

17. "Trends in the Electronics Industry in China," *Electronics Association of Japan*, April 1994, in *JPRS-CST*-95-001; also see "Shougang-NEC Electronic Company Goes into Limited Production," *Beijing Ribao*, November 23, 1994, in *FBIS-CHI*-94-234.

18. *Zhongguo Dianzi Bao*, March 26, 1996, p. 1, in *FBIS Science and Technology Perspectives*, Vol. 11, No. 3, June 1996.

19. Han Yuqi, "Nation Faces Challenge of MEMS Technology," *Keji Ribao*, November 6, 1996, in *FBIS-CST*-97-001. Leading figures in political and technical effort include Li Zhijian, Zhou Zhaoying, Wang Daheng, Zhang Wei, Tang Jiuhua, and Ding Henggao. For an AMS overview of nanotechnology, see Sun Bailin, "Nanotechnological Weapons Will Appear on the Battlefield: Prospects for Military Use of Micro-Scale Electromechanical Apparatuses," *Guofang*, June 15, 1996, pp. 39-40, in *FBIS-CHI*-96-151. A breakthrough in nanoscale optoelectronic measurement devices applicable to adaptive optics is discussed in "New Breakthrough in Nanoscale Optoelectronic Measurement Technology," *Keji Ribao*, March 28, 1997, p. 1, in *FBIS-CHI*-97-010.

20. See for example, Wang Qingsong, *Xiandai Junyong Gaojishu (Modern Military-Use High Technology)*, Beijing: AMS Press, 1993, pp. 251-254; Li, pp. 92-95; Zhu Youwen, Feng Yi, and Xu Dechi, *Gaojishu Tiaojianxia de Xinxizhan (Information War Under High Tech Conditions)*, Beijing: AMS Press, 1994, pp. 99-101, 163-167, and 270-280; and Wang Pufeng, pp. 113-116. Also see Wan Xuying, "Lun Zhanshu Daodan Weixie Yu Duikang Cuoshi" ("Discussion of the Tactical Missile Threat and Countermeasures"), *Zhongguo Hangtian*, November 1992, pp. 30-32.

21. PLA Deputy Chief of Staff Lieutenant General Xiong Guangkai is a key figure who oversees the PLA's entire intelligence apparatus. His portfolio also includes oversight of the PLA's foreign military-to-military relationships.

22. Nicholas Eftimiades, *Chinese Intelligence Operations*, Annapolis: Naval Institute Press, 1994, pp. 78-89.

23. Ping Kefu, "Capabilities of The GSD Third Department in Technical Intelligence," *East Asian Diplomacy and Defense Review*, 96 (5), p. 6.

24. Information on China's SIGINT apparatus drawn from Desmond Ball, "Signals Intelligence in China," *Jane's Intelligence Review*, August 1, 1995, pp. 365-375; and Robert Karniol, "China Sets Up Border SIGINT Bases In Laos," *Jane's Defense Weekly*, November 19, 1994, p. 5.

25. Ping Kefu, p. 6.

26. Zhou Lanying, "Overview of Signal Processing for Radar Countermeasure Information Reconnaissance," *Dianzi Duikang*, 1997, Vol. 2, pp. 9-14, in *CAMA*, Vol. 4, No.5; the KZ900 ELINT Pod was marketed in a brochure at the November 1998 Zhuhai Airshow.

27. Zhou Enlai directed initiation of the 701 Program and its launch vehicle, the FB-1, on August 14, 1969. The satellites operated in an inclination of 69-70 degrees, a perigee of 186 kilometers, an apogee of 464 kilometers, and a period of 90 minutes. The apogee on the last technical experiment satellite in August 1976 was raised to 2,145 kilometers. The August 1976 satellite decayed from orbit on November 25, 1978. China's technical experiment satellite of the 1970s is the only space platform not discussed in detail in official histories. ELINT satellite development is likely one of the most sensitive programs within the defense industry and PLA. One can not rule out the possibility that China has masked or will mask the presence of ELINT packages piggybacked onto other satellites. See *China Today: Space Industry*, pp. 162-170; "Summary of Launches in China," *Zhongguo Hangtian*, September 1996, p. 29; "Capabilities of The GSD Third Department in Technical Intelligence," p. 6; Liao Yuanshou, "Development of Satellite-borne Precision Direction Finding Antenna Array," *Dianzi Duikang Jishu* (Electronic Countermeasure Technology), March 1995, pp. 19-26, in *CAMA*, Vol. 2, No. 5. Liao is from the Southwest Institute of Electronic Equipment (SWIEE), also known as the 29th Research Institute, China's premier entity engaged in radar electronic support measures and ECM equipment; for another SWIEE piece on space ELINT, see Li Yudong, "Satellite-Borne Radar Reconnaissance," *Dianzizhan Xinjishu Yu Qingbao Gaige Wenzhai*, October 1995, pp. 126-133, in *CAMA*, Vol. 3, No. 6. For specific reference to SWIEE R&D on space-based electronic reconnaissance, see *China Today: Defense Science and Technology*, p. 742. Also see Yuan Xiaokang, "Satellite Electronic Reconnaissance, Antijamming," *Shanghai Hangtian*, October 9, 1996, pp. 32-37, in *FBIS-CST*-97-011; and Yuan Xiaokang, "Some Problems of Space Electronic Reconnaissance," *Hangtian Dianzi Duikang*, March 1996, pp. 1-5, in *CAMA*, Vol. 3, No. 4. Yuan is a key engineer involved in the space ELINT development from the SAST 509th Research Institute (Shanghai Institute of Satellite Engineering). Also see Li Ming, "Satellite-Borne Reconnaissance Antennas," *Hangtian Dianzi Duikang*, April 1997, pp. 38-41, in *CAMA*, Vol. 5, No. 2. Li is from the Nanjing Institute of Electronic Equipment.

28. Jeffery Richelson, "Navy Says China Poised to Close Space-Intel Gap," in *Defense Week*, February 24, 1997, p .9.

29. *Ibid.*

30. "Woguo Chenggong Fashe Yike Weixing" ("China Successfully Launches A Satellite"), *Zhongguo Hangtian*, November 1996, p. 10.

31. Bill Gertz, "Crowding In On the High Ground," *Air Force*, April 1997, pp. 18-22; and Covalt, pp. 22-23.

32. China's remote sensing program is funded at least in part by the 863 Program, specifically the 863-308 project. Hong Mei, "Tactical Application Satellite Imagery System," *Hangtian Fanhui Yu Yaogan (Spacecraft Recovery and Remote Sensing)*, 1995, Vol. 16, No. 1, pp. 30-33, in *CAMA*, Vol. 2, No. 5. A 700 kilometer optimizes coverage at the expense of resolution—a lower orbit naturally will increase the resolution. An advanced imaging system, perhaps the FSW-3, could be launched as early as 1999. See "China To Launch Ten More Satellites by 2000," *Xinhua*, February 22, 1998, in *FBIS-CHI*-98-053.

33. Jean Etienne, "Les Nouveaux Projets de L'Asie Spatiale," in *Space News*, No. 110, November 4, 1996, at http://www.sat-net.com/space-news. Also see Chou Kuan-wu, "China's Reconnaissance Satellites," *Kuang Chiao Ching*, March 16, 1998, pp. 36-40, in *FBIS-CHI*-98-098.

34. A broad survey of China's aerospace literature indicates the Second Department's Technology Bureau is a leading PLA proponent of photoreconnaissance systems. Leading Technology Bureau figures include Zhang Wanzeng.

35. Brochure, *National Remote Sensing Centre of China*, p. 4, no date.

36. Christian Lardier, "SPOT Image Pursues Changing Market," in *Cosmos/Aviation International*, February 14, 1997, p. 38, in *FBIS-EST*-97-005. A February 1997 Sino-French agreement also spells out cooperation in small satellite development, associated launch vehicles, joint development of satellite imagery systems.

37. "Woguo Jiang Zengshou Wuke Weixing de Guance Shouju" ("China Will Receive Five Sources of Surveying Data"), *Zhongguo Hangtian Bao (China Space News)*, January 5, 1997, p. 1; Fan Xizhe, "Future Importance of Information," in *Zhongguo Kexue Bao (China Science News)*, December 16, 1996, in *FBIS-CST*-97-003. Also see announcement by SSTC Vice Minister Xu Guanhua, Xinhua News Release, in *FBIS-CHI*-96-247. Even previous LANDSAT and SPOT imagery, with a resolution of 10-30 meters has significant military uses despite its poor resolution. For example, the U.S. Defense Mapping Agency used SPOT and LANDSAT imagery during the Gulf War to

update digital maps used by troops, strike mission planning, and bomb damage assessment.

38. Hiroshi Yuasa, "Russia Offering Spy Satellite Service to PRC," *Sankei Shimbun*, October 12, 1997, p. 1, in *FBIS-EAS*-97-286.

39. "China To Use Italian Software to Interpret Imagery," *Space News*, March 2-8, 1992, p. 23. Two other Chinese organizations involved in the project include China's Research Institute for Surveying and Mapping and the National Laboratory of Resources and Environmental Information Systems. Peng Yiqi, a senior engineer at the National Remote Sensing Center, led the Chinese negotiations.

40. "Woguo Jiang Zengshou Wuke Weixing de Guance Shouju" ("China Will Receive Five Sources of Surveying Data"), p. 1; Fan Xizhe, "Future Importance of Information," *Zhongguo Kexue Bao*, December 16, 1996, in *FBIS-CST*-97-003. Also see announcement by SSTC Vice Minister Xu Guanhua, Xinhua News Release. Twelve subscribers will be able to link into the stored images. Also see Chiu Fangying, "China's Remote Sensing Image Digitization Equipment Meets Advanced International Standards," *Keji Ribao*, October 26, 1995, p. 1, in *FBIS-CST*-96-002.

41. "Wo Weixing Yaogan Tuxiang Shuzihua Shebei Shijie Lingxian" ("Our Satellite Remote Sensing Digitized Imagery Equipment Leads the World"), *Zhongguo Hangtian*, January 1996, p. 39; also see "China's Satellite Remote Sensing Image Digitization Equipment Meets Advanced International Standards," *Keji Ribao*, October 26, 1995, in *FBIS-CST*-96-002.

42. Xu Guanhua and Guo Huadong, "Progress, Mission of Remote Sensing Research," *Yaogan Kexue Xin Jinzhan* (New Progress of Remote Sensing Science), April 1995, in *FBIS-CST*-96-002.

43. Long Zhihao, "Leida Weixing de Yingyong" ("Applications of Radar Satellites"), *Zhongguo Hangtian*, November 1991, pp. 29-31; Zhang Wanzeng, "Weixing Hecheng Kongjing Chengxiang Leida de Tedian Jiqi Zai Junshi Zhenchazhong de Yingyong" ("Applications and Characteristics of Satellite SAR for Military Reconnaissance"), *Zhongguo Hangtian*, November 1993, pp. 20-22. Zhang Wanzeng is assigned to the PLA GSD Second Department's Techology Bureau.

44. Huang Weigen, Zhou Changbao, and Wan Zhongling, "Woguo Xingzai SAR Haiyang Yingyong de Xianzhuang yu Xuqiu" ("Current State and Requirements of China's Satellite-borne SAR for Maritime Applications"), in *Zhongguo Hangtian*, December 1997, pp. 5-9.

45. Data collected by China's airborne SAR remote sensing platform can be transmitted real time to a ground station that is within 120 kilometers of the aircraft. A tactical ground processing system equipped with a VSAT terminal can then transmit the data to a command center. See "Remote Sensing Technical Systems for Reducing Flood Disasters," *Yaogan Kexue Xin Jinzhan*, April 1995, in *FBIS-CST*-96-002; and Xu Guanhua and Guo Huadong.

46. It is not clear if an actual contract has been signed or a Canadian export license issued. Zhong Bu, "Spar Commissions To Build Satellites for China," *China Daily (Business Weekly)*, July 14-20, 1996, in *FBIS-CHI*-96-136; Press releases, "China Signs Radarsat Data Reception and Distribution Agreement," *Spar Aerospace Limited*, April 25, 1996; Liu Wei, "Spar Intends to Join Hands With CASC," *Zhongguo Hangtian*, June 1996, in *FBIS-CST*-96-015; and "Spar Clarifies China Satellite Reports," *Spar Aerospace Limited*, July 15, 1996.

47. Steve Berner, "Proliferation of Satellite Imaging Capabilities: Developments and Implications," in *Fighting Proliferation: New Concerns for the Nineties*, Henry Sokolski, ed., Air University Press, 1996, pp. 95-128.

48. Wang Wei, "State S&T Organs Approve Design of Spaceborne Synthetic Aperture Radar," *Zhongguo Kexue Bao (China Science News)*, May 3, 1995, in *FBIS-CST*-95-010; "Woguo Xingzai Hecheng Kongjing Leida Yingyong Yanjiu Qude Zhongda Jinzhan" ("China's Satellite SAR Applied Research Achieves Tremendous Advances"), *Zhongguo Hangtian*, February 1996, p. 16; "Spaceborne-SAR Modern Information Technology Highlighted," *Zhongguo Kexue Bao*, September 20, 1996, p. 4, in *FBIS-CST*-96-020; Yuan Xiaokang, "High Speed Data Transmission of Satellite-borne SAR," *Zhidao Yu Yinxin* (Guidance and Detonators), 1995(4), pp. 8-14. (509th RI), in *CAMA*, Vol. 3, No. 4. Also see Yuan Xiaokang, "Performance Parameters and Design Requirements of Satellite SAR," *in Shanghai Hangtian*, 1996 (3), pp. 12-18; Long Zhihao, p. 29; Li Yudong, pp. 126-133. Li is from Xinan Dianzi Shebei Yanjiusuo (SWIEE). For comments on preliminary research on the second generation SAR satellite, see "China's Microwave-Imaging Radar Systems Engineering Highlighted," *Zhongguo Kexue Bao*, September 20, 1996, p. 4, in *FBIS-CST*-96-020.

49. Chang Jijun, "Remote Sensing Image Data Compression and Its Performance Evaluation," *Kongjian Jishu Qingbao Yanjiu*, July 1994, pp. 37-54, in *CAMA*, Vol. 1, No. 6.

50. Zhang Wanbin, "Spaceflight Development Strategy: Mid-Long Term Development Strategy, *Zhongguo Keji Luntan (Forum on Science*

and Technology), November 1992, pp. 9-12, in *JPRS-CST*-93-002; and Cheng Yuejin, "Information Transmission System of Data Relay Satellites," *Space Technology Information Research*, 1993, summarized in *CAMA*, 1994, Vol. 1, No. 6.

51. Wang Chunyuan, *China's Space Industry and Its Strategy of International Cooperation*, Stanford: Stanford University, July 1996, p. 4. Lieutenant Colonel Wang serves on the senior staff of COSTIND's foreign affairs bureau.

52. Wang Mingyuan, "Mobile Remote Sensing Ground Stations," *Kongjian Dianzi Jishu (Space Electronic Technology)*, 1997 (2), pp. 32-37, in *CAMA*, 1997, Vol. 4, No. 6.

53. Zhang Yinlong, "Xian Satellite Control Center, China Satellite TT&C Network," *Zhongguo Hangtian*, October 1991, pp. 44-47, in *JPRS-CST*-92-006; deployment of a space based tracking system is explictly listed in CASC's long-range development plan. Also see Dong Guangliang, "A Computing Method for Determining Satellite-to-Satellite Tracking Visual Interval," *Feixingqi Cekong Jishu* (Flight Tracking Technology), February 1994, pp. 20-25, in *CAMA*, Vol. 1, No. 5.

54. DoD Report to Congress, Pursuant to Section 1226 of the FY98 National Defense Authorization Act.

55. See Zhang Zhe, "Development of Over-the-Horizon Sensor Technology," *Zhongguo Hangtian*, August 1991.

56. This site, allegedly visible on the road from Beijing to Zhangjiakou, may be inactive.

57. *China Today: Defense Science and Technology*, pp. 728-729.

58. ASN-206 Brochure, Xian Aisheng Jishu Jituan; Li Youhao and Yang Changhua, "Development and Successful Flight Test of China's First High Altitude Photographic Surveillance UAV," unpublished BUAA paper, April 1994; also see Ji Xiumin, "Modern UAVs Profiled," *Hangkong Zhishi*, October 1997, pp. 41-47, in *FBIS-CHI*-98-083.

59. Wang Yongsheng, "Design and Implementation of a Navigation System for UAVs Using GPS," *Xibei Gongye Daxue Xuebao*, 1996, Vol. 14, No. 3, in *CAMA*, Vol. 3, No. 6; and Zhang Zhong'an and Wang Lue, "Contour Design for a Low Radar Cross-Section Unmanned Aerial Vehicle," *Yuhang Xuebao (Journal of Astronautics)*, February 1, 1996, pp. 43-47, in *FBIS-CST*-96-017.

60. DoD 1998 Report to Congress, pursuant to Section 1226 of the FY98 National Defense Authorization Act.

61. AT&T News Release, "AT&T Signs Broad Agreements for Telecom Infrastructure in China," April 28, 1994.

62. Hao Yunpeng, *China's Telecommunications: Present and Future*, Stanford: Center for International Security and Arms Control, June 1997. For China's long-range plans, see "Xinxi Gaosu Gonglu de Fazhan he Duice Yanjiu," ("Development and Policy Research on Information High Speed Highway"), *Hangtian Jishu yu Minpin (Space Technology and Civil Products)*, April 1996, pp. 21-26.

63. Wang Shengrong, "The Information Superhighway and Military Communications," *Guofang* (National Defense), February 15, 1995, in *FBIS-CHI*-95-100; also see Zhang Xiaojun, October 15, 1995.

64. Alexandra Rehak and John Wang, "On the Fast Track," *The China Business Review*, March-April 1996, pp. 8-13; PLA involvement was reported when news of Liantong's formation first broke in early 1993. Since then, however, Liantong has downplayed PLA involvement. See Geoffrey Crothall, "New Telephone Network Planned," *South China Morning Post*, February 18, 1993. One other source which commented on PLA communications network being linked into an integrated national network is a 1994 interview with Liantong director Zhao Weichen, in Xie Minggan, "Communications Company Chief Interviewed," *Ching Chi Tao Pao (Economic Reporter)*, August 29, 1994, in *JPRS-CAR*-94-053.

65. Author's Fall 1993 visit to CAEIT, also known as China Electronics Corporation, Systems Engineering General Department. As of 1993, CAEIT was directed by Madame Hu Bangde.

66. "Fiber Optic Communications Technology," *Zhongguo Dianzi Bao*, December 1, 1995, p. 3, in *FBIS-CST*-96-005; Liu Dongsheng, "Telecommunications: Greater Sensitivity Achieved," *Jiefangjun Bao*, September 8, 1997, p. 5, in *FBIS-CHI*-97-287.

67. Tang Shuhai and Guan Ke, "All-Army Public Data Exchange Network Takes Initial Shape," *Jiefangjun Bao*, September 18, 1995, in *FBIS-CHI*-95-230; and Liu Dongsheng, "Telecommunications: Greater Sensitivity Achieved," *Jiefangjun Bao*, September 8, 1997, p. 5, in *FBIS-CHI*-97-287.

68. For example, in the United States, over 95 percent of DoD and the intelligence community voice and data traffic uses the public telephone system, and this amount is likely to increase.

69. Tan Keyang, "Defense Military Computer Network Interconnects PLA Army, Navy, and Air Force," *Jisuanji Shijie*, August 11, 1997, No. 30, p. 1, in *FBIS-CHI*-97-324.

70. See Dennis Blasko, Ray Lawlor, John Corbett, Mark Stokes, and Chris Kapellas, *China's Defense-Industrial Trading Organizations*, Defense Intelligence Reference Document, October 1995; and Foo Choy Peng, "Dutch Firm Forges Telecom Partnership with PLA Unit," *South China Morning Post*, February 27, 1998, p. 27, in *FBIS-CHI*-98-058. The other two networks include that of MPT and Liantong. CESEC has recently entered into a joint venture with a Dutch company, KPN.

71. Author's January 1993 meeting with BISE representatives; and *Directory of PRC Military Personalities*, USDLO Hong Kong, October 1996, p. 13. The GSD Telecommunications Department is directed by Major General Yuan Banggen. Until his recent retirement, the individual most involved in COSTIND's C^4I systems integration effort was Chang Mengxiong, who has emerged to become a leading advocate for full exploitation of the RMA. At least one GSD subordinated research institute, probably under the Telecommunications Department, the 61st Institute, has communications R&D responsibilities.

72. See, for example, "Communications Project Approved," in *Qinghai Ribao*, July 21, 1993, which outlined a joint MPT-Second Artillery (80306 Unit in Xining) project to install a 480 channel digital microwave link between Xining and Golmud.

73. China's C^2 apparatus is detailed in Michael D. Swaine, *The Military and Political Succession in China*, RAND Corporation, 1992, pp. 122-123.

74. You Ji, "High Tech Shift For China's Military," *Asian Defense Journal*, September 1995, pp. 4-10; Viacheslav A. Frolov, "China's Armed Forces Prepare for High-Tech Warfare," *Defense & Foreign Affairs: Strategic Policy*, January 1998, pp. 12-13; and Nanjing Research of Electronics Technology (NRIET) brochure on 3 Meter Offset-Fed SATCOM antenna.

75. Hao Yunpeng, *China's Telecommunications: Present and Future*, Stanford: Center for International Security and Arms Control, June 1997, p. 1.

76. Wang Boyi, "State 863 Program's Communications Technology Area's Ninth FYP Research Projects Detailed," *Dianxin Jishu*, June 10, 1997, pp. 4-6, in *FBIS-CST*-97-010; Wei Leping, "Higher Speed, Longer Wavelength, Multiple Wavelength—Major Trends of Optical Fiber

Systems For Long Haul Transmission in China," China Telecommunications Construction, July 3, 1996, pp. 20-26, in *FBIS-CST*-96-009; "Advanced Telecommunications Equipment Joint Venture Set Up," *Xinhua*, February 2, 1996. Current SDH fiber-optic cable systems have transmission rates ranging from 2.5-10 gigabits per second. After 2002, according to a DoD 1998 report to Congress on PLA military capabilities, China's fiber-optic network will be completely composed of SDH.

77. For background on some of these communications technologies, see *New World Vistas: Air and Space Power for the 21st Century, Information Applications Volume*, 1995.

78. Xu Dianlong, "Beijing Reports 8 Dec Satellite Launch," *Xinhua*, December 9, 1997, in *FBIS-CHI*-97-349.

79. Zhang Xinzhai, "Achievements, Future Development of China's Space Technology," *Aerospace China*, June 1996, in *FBIS-CST*-96-015. Also see "China To Launch Ten More Satellites by 2000," *Xinhua*, February 22, 1998, in *FBIS-CHI*-98-053.

80. Zhou Quan, "Spread Spectrum Techniques for Satellite Communications," *Kongjian Dianzi Jishu*, 1997, (2), pp. 5-9, in *CAMA*, Vol. 4, No.6; and Xiao Kai, "Synchronization Concepts for DH/DS Spread Spectrum Receivers," *Kongjian Dianzi Jishu*, 1997 (2), pp. 10-15, in *CAMA*, Vol. 4, No. 6. Zhou and Xiao are from the Xian Institute for Space Radio Technology, the primary organization responsible for satellite transponders and remote sensing downlinking technology.

81. "Company Plans Asian Satellite," *Space News*, June 1-7, 1992. COSTIND's participation as an investor provides the necessary bureaucratic leverage for the PLA to use transponder space should the need arise. The PLA generally relies on domestic satellites, such as the DFH-3, for its communications requirements.

82. Ji Hongguang, "MPT Buys CHINASAT-8 From Space Systems Loral," *Keji Ribao*, March 24, 1997, in *FBIS-CHI*-97-010.

83. China Orient is paying $100 million for the satellite including ground support equipment. "Long March to Launch Satellite," *China Daily*, August 29, 1996, in *FBIS-CHI*-96-169. MPT-run satellite TT&C ground stations in Beijing monitor operating conditions of the satellites. However, COSTIND's Xian Satellite Control Center will exercise satellite station keeping responsibilities.

84. Berner, p. 110. The specific device is the DF50 momentum wheel, which can also be used in remote sensing platforms. A viable

attitude control system remains the most vexing problem for China's space industry. Attitude control problems lie at the root of DFH-3 failures, and, based on author's June 1998 discussions with China's weather officials, the FY-2 weather satellite system.

85. "The Iridium System," *Iridium Homepage, www.iridium.com.*

86. Chen Shupeng, "China: Geomantics, Regional Sustainable Development," April 1995, in *FBIS-CST*-96-02.

87. "Singapore Firms Agree on Satellite Communications," *Xinhua*, December 28, 1995, in *FBIS-CHI*-96-002; and Min Changning, "Concepts or China, Asia-Pacific Mobile Telecom Satellite," *Zhongguo Hangtian*, November 1994, in *JPRS-CST*-95-003.

88. Ma Rushan, "Status and Development of China's Golden Bridge Network," *Jisuanji Shijie (Computerworld)*, September 13, 1995, pp. 195-197, in *FBIS-CST*-96-001.

89. "National Toll-Telephone Network Intelligent Network Project Gears Up," *Dianxin Jishu*, June 1996, p. 48, in *FBIS-CST*-96-017; "Shanghai Bell's R&D Budget Reaches 300 Million Yuan," *Zhongguo Dianzi Bao*, September 26, 1995, p. 1, in *FBIS-CST*-96-001; "BTA, Guangdong Nortel Sign Contract for Development of Intelligent Network," *Keji Ribao*, May 29, 1996, p. 5, in *FBIS-CST*-96-013; Wu Jichuan, "China's Telecommunications Striding Into the Ninth Five-Year Plan," *Dianxin Jishu*, March 27, 1996, "Into the Ninth Five-Year Plan," *Dianxin Jishu*, March 27, 1996, pp. 3-5, in *FBIS-CST*-96-005, Gao Hui, "China's First Set of High Intelligence Network Systems Has Been Tested and Accepted," *Jiefangjun Bao*, December 5, 1997, p. 1, in *FBIS-CHI*-97-357; and "China's Intelligent Networks," *Keji Ribao*, May 29, 1996, p. 5, in *Science and Technology Perspectives*, Vol. 11, No. 4, August 1996, p. 3-5, in *FBIS-CST*-96-005.

90. A significant proportion of technical writings on military software development originate from BISE. Examples include He Xingui, "Activeness of Software and Its Military Application," *China Computer Society Meeting Abstracts*, May 1994, pp. 1-7, in *CAMA*, Vol. 1, No. 6; and Liu Shuming, "Quality Measurement Model and Testing of Military Software," *China Computer Society Meeting Abstracts*, May 1994, pp. 131-135, in *CAMA*, Vol. 1, No.6. Both authors are from BISE.

91. Liang Zhenxing, "Information Warfare: Major Influence on China's Defense Modernization," in *Jisuanji Shijie (China Computerworld)*, April 8, 1996, in *FBIS-CST*-96-009; also see Liang Zhenxing and Lan Guoxing, "Summary of Integrated Electronic

Information (C^4I) System Development," *Jisuanji Shijie*, February 3, 1997, No. 5, pp. 94-95. Liang Zhenxing is from the Beijing Institute of Systems Engineering, a COSTIND entity which plays a central role in the development of China's national C^4I network.

92. Li Guangru, "Zou Jiahua Addresses Information Leading Group's Inaugural," *Xinhua*, May 27, 1996, in *FBIS-CHI*-96-119.

93. Chen Zhujia, "Marching Towards the Commanding Height of Science and Technology," *Renmin Ribao*, May 29, 1995, in *FBIS-CHI*-95-117; and Qin Chun, "Higher Speed Large Parallel Computer System To Be Built," *Xinhua*, January 31, 1996, in *FBIS-CHI*-96-024.

94. Qian Weihua, "Institute Develops Advanced Space Computers," *Wenhui Bao*, February 5, 1991, in *FBIS-CHI*-91-029. Institutes involved in microcomputer development include the 771st Institute in Lishan and Huadong Computer Technology Institute in Shanghai.

95. Tan Keyang, "YH Super Simulation Computer, YH Intelligent Computer Systems Certified," *Jisuanji Shijie*, October 21, 1996, p. 1, in *FBIS-CST*-97-002.

96. Caption under picture of Russian Defense Minister Touring Beijing Simulation Center, *Zhongguo Hangtian Bao*, April 23, 1997, p. 1.

97. Wang Chengwei, "Review and Near Term Prospects for Intelligent Computer Systems," *Scientia Sinica (Science in China)*, Vol. 38, September 1995, pp. i-iv, in *FBIS-CST*-96-009.

98. "IBM Chief Says China Poised To Lead in Networking Technology," *AT&T Homepage*, nando.net.

99. Liang Baichuan, "Experimental Studies on Computer Virus Injecting Technology Through Radiation," *Huoli Yu Zhihui Kongzhi (Firepower and Command Control)*, Vol. 20, No. 4, 1995, in *CAMA*, Vol. 3, No. 3, 1996. The author conducts research at the Xian University of Electronic Science and Technology.

100. There is still a gap in knowledge about the GSD Fourth Department which is rapidly becoming one of the most significant GSD entities. One of the first references to outline GSD Fourth Department missions is Ping Kefu, p. 6. While the GSD Third Department has the communications intelligence portfolio, the Fourth Department is responsible for ELINT and ECM operations.

101. Zhang Youcai, "Denglu Zhanyi Dianzi Duikang Zuozhan Zhidao de Jige Wenti" ("Some Questions Surrounding Theater ECM Operational Principles"), in *Zuozhan Zhihui Yanjiu (Research on Operational; Command and Control)*, Beijing: National Defense University Press, January 1997, pp. 327-333. Based on his rank, MGen Zhang Youcai is likely director or deputy director of the GSD Fourth Department.

102. Mei Zhongtang, "Electronics: Building Up the Republic's 'Information Great Wall'," *Jiefangjun Bao*, September 11, 1997, p. 5, in *FBIS-CHI*-97-287.

103. Li Yongqiang, "Applications of UAVs in Electronic Warfare," *Dianzi Duikang Jishu*, February 1995, pp. 15-21, in *CAMA*, Vol. 2, No. 5. Li is from the Southwest Institute of Electronic Equipment (SWIEE); also see Li Jiaxiang, "Development of UAVs for Electronic Warfare," *Hangtian Dianzi Duikang*, March 1994, pp. 43-47, in *CAMA*, Vol. 1, No. 6. Besides jammers, SWIEE carries out R&D into airborne and space-based radar reconnaissance platforms.

104. Among numerous references, see Zhang Shenghai, "Research on ECM Equipment Science," *Proceedings of New Electronic Warfare Technology and Intelligence Reform Conference*, October 1995, pp. 18-21, in *CAMA*, Vol. 3, No. 6.

105. Wu Jinliang, "Range Testing of Satellite Communication Countermeasures," *Proceedings of New Electronic Warfare Technology and Intelligence Reform Conference*, October 1995, in *CAMA*, 1996, Vol. 3, No. 6. Wu Jinliang is from the PLA 89823 unit, most likely a COSTIND-subordinated entity. The Shijiazhuang Institute of Telecommunications and Tracking Technology (54th Research Institute) is one entity which has has conducted R&D on SATCOM jammers. See Zhu Qinghou, "Satellite Communications Countermeasures in Information Warfare," *Hangtian Dianzi Duikang*, 1997, Vol. 1, pp. 23-26, in *CAMA*, Vol. 4, No. 4; Zhu Qinghou, "Research On Jamming Methodologies of Military Multichannel Satellite Transponders," *Hangtian Dianzi Duikang*, March 1997, pp. 31-34, in *CAMA*, Vol. 5, No. 2. Zhu is from the Shijiazhuang Institute of Communications, Tracking, and Control Technology.

106. Chen Ning and Zhang Jieru, "Research Into Synthetic Aperture Radar Jamming Technology," *Hangtian Dianzi Duikang*, April 1997, pp. 45-48, in *CAMA*, Vol. 5, No. 2. One May 1998 source with direct access to commercial SAR imaging services claims that the Chinese may have attempted to jam a foreign SAR satellite as it passed over a 1996 PLA exercise in Fujian province.

107. Liang Baichuan, "The Jamming of the Global Positioning System," *Shanghai Hangtian*, March 1996, pp. 35-40, summarized in *CAMA*, Vol. 3, No. 5; and author's unpublished study, "PLA Strategic Modernization: The Role of China's Space and Missile Industry," September 1995. Two institutes carrying out research into GPS jammers include Jiangnan Electronic Communications Research Institute in Guizhou province (061 Base) and Xian University of Electronic Science and Technology (*Xidian*).

108. Liang Baichuan, "Technologies for Jamming AWACS in China's National Air Defense System," *Shanghai Hangtian*, 1994, 11(3), in *CAMA*, Vol 1. No. 5; and Liang Baichuan, "Main Lobe Jamming Against Phased Array Airborne Early Warning Radars," *Shanghai Hangtian*, 1997, 14(6), pp. 12-15, in *CAMA*, Vol. 5. Liang Baichuan, from the Xian University of Electronic Science and Technology (*Xidian*), is one of China's most prolific writers on information operations. Xidian, reportedly one of China's highest level military communications institutes, has conducted computer simulation on AEW jammers. Also see Tao Benren, "Ground-based Reconnaissance Systems for Countering Airborne Warning and Jamming Aircraft," in *Hangtian Dianzi Duikang (Space Electronic Countermeasures)*, 1995, (2), pp. 40-44, in *CAMA*, Vol. 2., No. 5. Tao is an electronic warfare specialist from the Shanghai Institute for Electromechanical Engineering. On electronic combat against JTIDS, see Xu Xiutao, "Countermeasures Against Joint Tactical Information Distribution System," in *Dianzi Duikang*, February 1995, pp. 1-7, in *CAMA*, Vol. 2, No. 5. Xu is from the Jiangnan Institute of Electronic Communication, the same institute which is carrying out R&D into GPS jammers.

109. Prasun Sengupta, "Asia-Pacific Enters the Electronic Warfare Arena," *Aerospace*, June 1992, pp. 28-30.

110. A flux compression generator (FCG) is the most mature technology applicable to HPM bomb designs. Lin Zheng, "New Advances in Electronic Warfare," in *Proceedings of '96 Conference Sponsored By Journal "Huoli Yu Zhihui Congzhi,"* October 1996, pp. 16-21, in *FBIS-CST-97-012*; Zhang Zhenzhou, "Longitudinal Tranmission of Exploding Electromagnetic Waves," in *Xiandai Fangyu Jishu*, April 1995, pp. 47-58, in *CAMA*, Vol. 2, No. 5. Also see Qin Zhiyuan, "HPM Weapons in Tomorrow's War," paper presented at 1997 COSTIND-sponsored international conference on RMA. See Appendix V for more details.

111. Li Guangpu, "Long Range Multipurpose Air-to-Air Missile: Most Effective Weapon to Strike Radiating High Value Airborne Assets," *Yuancheng Kongkong Daodan Zhuanji (Long Range Air-to-Air*

Missile Special Collection), June 1996, pp. 1-13, in *CAMA*, Vol. 4, No. 2. Li is from the Shanghai Academy of Space Technology's primary missile design institute, the Shanghai Institute of Electromechanical Engineering (SIEE). Also see Xing Xiaolan (PLAAF 8th Research Institute, responsible for AAM program management), "Development of Anti-AEW Aircraft Weapon System," *Yuancheng Kongkong Daodan Zhuanji,* June 1996, pp. 14-21, in *CAMA*, Vol. 4, No. 2; Zhang Dechang (Shanghai Institute of Engineering Technology Information), "Development Trends of Long Range Air-to-Air Missiles," *Yuancheng Kongkong Daodan Zhuanji,* June 1996, pp. 22-43; Tao Benren (SIEE), "Characteristics of AWACS and Airborne Jammers," *Yuancheng Kongkong Daodan Zhuanji,* June 1996, pp. 54-57; Zhang Wangsheng (SIEE), "Guidance and Control Technology of Long Range Air-to-Air Missiles," *Yuancheng Kongkong Daodan Zhuanji,* June 1996, pp. 58-65; Wang Chaoqun (SIEE), "Application of Passive Microwave/Infrared Imaging Dual Mode Complex Homing Technology For Long Range Multipurpose Air-to-Air Missiles," *Yuancheng Kongkong Daodan Zhuanji,* June 1996, pp. 66-72, in *CAMA*, Vol. 4, No. 2; Hu Fuchang (Shanghai Institute of Radio Equipment), "Design of Passive Microwave Seeker," *Yuancheng Kongkong Daodan Zhuanji,* June 1996, pp. 73-81, in *CAMA*, Vol. 4, No. 2; Wu Youliang, "Airborne Passive Microwave Detection Positioning System," *Yuancheng Kongkong Daodan Zhuanji,* June 1996, pp. pp. 82-94, in *CAMA*, Vol. 4, No. 2; Min Bin, "Multiple-Start Pulsed Solid Rocket Motor," *Yuancheng Kongkong Daodan Zhuanji,* June 1996, pp. 95-101, in *CAMA*, Vol. 4, No. 2 (Min is from the Shanghai Xinli Power Equipment Institute); Zhang Wangsheng (SIEE), "Development of Long Range Multipurpose Air-to-Air Missile: Constitution of A Counterair Operational System," paper presented at China Space Society Conference on UAVs, November 1996, 11 pp.

112. Outlined in James Dunnigan, *Digital Soldiers: Evolution of High Tech Weaponry and Tomorrow's Battlefield*, New York: St. Martin's Press, 1996, p. 180.

113. 1998 China Precision Machinery Import and Export Corporation FT-2000 brochure.

114. Kang Hengzhen, "The Rise of Special Operations," *Jiefangjun Bao*, August 13, 1996, p. 6, in *FBIS-CHI*-96-187; and Zhou Mengwu, "Uncovering the Secrets of the Chinese Special Forces," *Junshi Wenzhai*, October 5, 1997, in *FBIS-CHI*-98-037.

115. Dai Shenglong and Shen Fuzhen, *Xinxizhan Yu Xinxi Anquan Zhanlue* (Information Warfare and Information Security Strategy), Beijing: Jincheng Publishing House, 1996, pp. 148-164. Also see Cai

Delu and Li Ruifeng, "Electronic Security Technologies and Standards," *Zhongguo Dianzi Bao*, January 21, 1997, p. 11, in *FBIS-CHI*-97-199. Dai is Director of the State Secrets Bureau. On the CAS INFOSEC Research Center, see Chen Xiechuan, "CAS Information Security Technology Engineering Research Center," *Zhongguo Kexue Bao*, August 30, 1996, p. 1, in *FBIS-CST*-96-019. On the National Key Laboratory, see Zhao Zhansheng, "Information Security and the Three Golden Projects," *Keji Daobao*, April 1995, pp. 29-32, in *FBIS-CST*-95-010.

116. Chou Hsi, "Exploration and Analysis of Military Computer Security and Virus Protection," *CONMILIT*, January 11, 1996, pp. 34-35, in *FBIS-CHI*-96-116; Gan Shaowu, "Computer Virus Offense Patterns, Methods Viewed," *Jiefangjun Bao*, July 30, 1996, in *FBIS-CHI*-97-107; Xu Renjun and Chen Xinzheng, "Computer Virus Weapons," *Guofang*, February 15, 1997, pp. 42-44, in *FBIS-CHI*-97-073; "Expert Suggests Building Information Security System," *Xinhua* news release, March 6, 1997, in *FBIS-CHI*-97-065; Xiao Zheng, "Security of Information Vital Next Step for China," *China Daily*, March 18, 1997, p. 4; and Zheng Qianli, "Building a Powerful 'Information Boundary'—Interview With Information Warfare Expert," *Keji Ribao*, April 30, 1997, p. 2, in *FBIS-CST*-97-012.

117. Sun Yougui, "A Look at Military Schools," *Military Prospect*, April 1998, pp. 31-38.

118. Mao Guanghong, "On Electromagnetic Management of the Modern Battlefield," *Jiefangjun Bao*, May 21, 1996, p. 6, in *FBIS-CHI*-96-134.

119. Zhang Jian, "Analysis of ECCM Principles of Spread Spectrum Unified Satellite Tracking, Telemetry, and Control Network," *Hangtian Dianzi Duikang*, April 1997, pp. 26-30, in *CAMA*, Vol. 5, No. 2. Zhang is from the China Academy of Engineering Physics' Electronic Engineering Institute. Also see Wei Chenxi, "ECCM Measures for Military Communications Satellites," *Hangtian Dianzi Duikang*, March 1997, pp. 31-34, in *CAMA*, Vol. 5, Vol. 1.

120. Qin Zhongping and Zhang Huanguo, "ALT: Algorithim for Attacking Cryptosystems," *Jisuanji Xuebao*, Vol. 20, No. 6, pp. 546-550, in *FBIS-CHI*-97-311; and Zhou Hong and Ling Xieting, "Encryption By Inverse Chaotic Systems," *Fudan Xuebao*, June 1997, Vol. 36, No. 3, pp. 301-308, in *FBIS-CHI*-97-281.

121. Sun Zi'an, "Strategies To Minimize High Tech Edge of Enemy," *Xiandai Bingqi* (*Modern Weaponry*), August 8, 1995, pp. 10-11, in

FBIS-CHI-96-036; Mi Zhenyu, "China's National Defense Development Concepts," in Michael Pillsbury, *Chinese Views of Future Warfare*, National Defense University Press, 1997; and Xu Xingju, "Tigao Woguo Dimian Fangkong Wuqi Dianzizhan Nengli" ("Raising China's Ground Based Air Defense Electronic Warfare Capability"), *Xiandai Bingqi*, August 1995; and Li Chunshan, "Introduction of National Military Standard 'Camouflage Requirements of Surface-to-Surface Missile Systems'," *Hangtian Biaojunhua*, May 1994, pp. 12-15, in *CAMA*, Vol. 2, No. 1; and Kang Qing, "Infrared Camouflage Mechanisms and Applications," *Hongwai Jishu* (Infrared Technology), January 1996, pp. 25-27, in *FBIS-CST*-96-016.

122. Kang Qing, "IR Stealth of Buried Targets," *Hongwai Jishu*, 1996, 18 (6), pp. 21-24, in *CAMA*, 1997, Vol. 4, No. 1. Kang is from the PLA Academy of Logistics Engineering.

123. See, for example, Huang Peikang and Yin Hongcheng, "A New Stealth Technique for Flight Vehicles," *Yuhang Xuebao*, 1994, Vol. 15, No. 2.

CHAPTER 4

DAWN OF A NEW AGE: CHINA'S LONG-RANGE PRECISION STRIKE CAPABILITIES

With a solid base of information, PLA planners hope someday to be able to mate sensors and strike assets into a real-time strategic reconnaissance/strike complex. The PLA hopes to develop a range of strike assets with which to exploit adversarial vulnerabilities on the ground, on the ocean, in space, and within the electromagnetic spectrum. Chinese strategists and planners have priority programs to develop a new generation of lethal, highly accurate, and survivable ballistic and cruise missiles, able to penetrate any theater or national missile defense system. Since 1991, the PLA and the space and missile industry have made conventionally armed surface-to-surface missiles and extended range cruise missiles key projects in their 15-year developmental plans.

Shortly after the Gulf War, the PLA expanded the role of its strategic rocket forces, known as the Second Artillery, as a conventional striking force. The new mission is the direct result of a dramatic increase in the accuracy of its ballistic missile force. Missiles are rapidly becoming the sole credible long-range firepower projection asset which the PLA has in its inventory, and this will remain likely true for the foreseeable future.[1] This chapter examines selected long-range strike assets which could allow the Chinese to achieve their limited objectives. An examination of China's cruise missile program is first explored, followed by an outline of its ballistic missile research and development, and concluded with an overview of China's strategic missile force, the Second Artillery.

Cruise Missiles.

Since the Gulf War and the increasing availability of technology, the Chinese defense industrial complex has decided to upgrade its aerodynamic missiles. Building upon a well-established foundation of air launched, ground launched, and submarine launched anti-ship missile technology, China is creating a new generation of cruise missiles able to penetrate defenses and strike critical targets with precision and increased firepower. Fielding of land attack cruise missiles will prompt expansion of missions of the PLA's Second Artillery and Navy.

The heart of China's aerodynamic missile development lies within CASC's Third Academy (*di san yanjiuyuan*, or *sanyuan* for short), headquartered in Yungang, southwest of Beijing. Over 14,500 technicians and workers ply their trade in ten research institutes and two major factories. Each of the ten research institutes focuses on specific subcomponents of a cruise missile, such as propulsion, inertial navigation, control systems, terminal guidance, and launching equipment. The Third Academy's Third Design Department, consisting of over 1,000 technicians, has systems engineering and design responsibilities.[2]

The Third Academy's cruise missile R&D effort began in the late 1950s. Intended as an anti-ship coastal defense system, Qian Xuesen led the industrial campaign. Based on an agreement signed in 1959, the Soviet Union provided prototypes and technical information. R&D and production facilities were established under the space and missile industry at Yungang near Beijing, in Hubei province (066 Base) and in Sichuan (062 Base). The aviation industry established an augmenting, and sometimes competing, center for anti-ship missile production at the Nanchang Aircraft Factory. With Soviet assistance, an anti-ship missile testing complex was established off the coast of Liaoxi, Liaoning province. Since 1959, over 10 variants of anti-ship missiles have been produced.[3]

The Gulf War provided a new impetus for cruise missile development. The PLA and the defense industries were awed by the performance of U.S. and allied land attack cruise missiles, especially the Tomahawk. Since then, the Third Academy has flooded the civilian and military leadership with feasibility studies on air, ground, and sea launched aerodynamic missiles. The PLA and the Third Academy have moved beyond the concept of utilizing cruise missiles for tactical coastal defense, to a more ambitious goal of fielding missiles which are able to strike strategic land- and sea-based targets.

The defense industrial complex strongly advocates increasing investment into cruise missiles. In an offense-defense cost comparison, some believe that developing an arsenal of cruise missiles can have a 9:1 ratio between developing, deploying, and maintaining cruise missiles over the cost of defending against them. As the president of the Third Academy has pointed out, the cost of producing cruise missiles is 20-30 percent less in China than it is in other countries.[4] A national program aimed developing cruise missiles will further foster technological spin-offs from work in inertial guidance, terrain matching, scene matching terminal guidance, turbojet and turbofan engines, supercritical bomb wings, electronic warfare, and concealment and counterconcealment technology. Third Academy representatives also assert that highly capable cruise missiles can be profitable by marketing their products overseas.[5]

Third Academy director Wang Jianmin has outlined what he considers to be priorities for future cruise missile development. First and foremost is a completely new generation of mid- to long-range cruise missiles included in the 2010 long-range plan. He outlined specific technologies, such as low smoke solid fueled motors, composite ramjet engines, IR imaging guidance, active/passive microwave/IR imaging guidance, and IR/laser imaging composite guidance, and new forms of ECCM and on board jammers, as top priorities in R&D efforts.[6]

Chinese analysts see three applications of cruise missiles. First, they clearly envision using cruise missiles in an offensive counterair role. Highly accurate missiles can offset an adversary's superiority in air forces by striking its air assets while they are concentrated on the ground or on a carrier. Other potential targets include runways, air operations centers, POL facilities, and quarters housing pilots. Such a targeting capability requires near real-time sources of information, a high degree of accuracy, and effective munitions.[7] The Third Academy has conducted in-depth targeting assessments of U.S. aircraft carriers, including fleet defense assets, characteristics of radar and infrared emissions, and electronic combat capabilities.[8]

Chinese strategists also discuss cruise missiles as a critical element in their quest for information dominance. Cruise missiles are necessary to take out an adversary's eyes and ears. This includes antiradiation missile (ARM) strikes against radars, land attack cruise missile attacks against high value reconnaissance assets on the ground, and strikes against C^4I nodes in order to paralyze command and control. China's defense industrial complex, specifically the Third Academy with support from the Harbin Institute of Technology, is aggressively pursuing deployment of a long-range antiradiation missile.[9] An ARM capability would be a significant leap in the PLA's ability to hit important targets protected by air defense assets. A final utility of cruise missiles is nuclear deterrence. There are clear indications that China will likely have a nuclear warhead sized for one of their cruise missiles and are seriously considering adding another leg to their nuclear force.[10]

Hoping to field a viable aerodynamic missile which is able to hit critical targets at sea and on land, the defense industrial complex is focusing resources on a few key areas potentially applicable to all variants under development.[11] First, the Third Academy is striving to reduce the radar cross section (*leida jiemian*) of its missiles, potentially requiring a new class of cruise missile. Specific methodologies include shaping of the airframe and application of the radar absorbing materials.[12] Reduced cross section is needed to minimize an adversary's ability to detect an incoming missile or at least reduce reaction time.

Engineers are also working on better propulsion systems which can increase the lethal range or speed of the cruise missile. Faster cruise missiles reduce an adversary's reaction time. In one of China's most significant aerospace programs, COSTIND and the Third Academy are designing a supersonic combustion ramjet engine (scramjet, or *chaoran chongya fadongji*) which can propel a missile at hypersonic speeds of between Mach 4-10.[13] Engineers are also working toward more efficient turbojet and turbofan engines and motors in order to significantly extend the range of its cruise missiles. The current range of China's antiship missiles is limited to around 200 kilometers. To be able to hit targets in Japan using a ground launched system, for instance, the Third Academy would have to produce a missile with an approximately 1,250-1,500 kilometer range (750 kilometers for Okinawa).[14]

The Third Academy is also capitalizing on many available technologies in order to improve the accuracy of its missile force. At least one of the Third Academy's requirements driving the next generation of cruise missiles is the ability to strike airfield runways with conventional munitions, a feat which requires tremendous accuracy. China's rapid rise in computing capabilities and growing availability of commercial satellite imagery provides increased opportunity to integrate terrain command guidance (TERCOM) and digital scene matching area correlator (DSMAC) into their navigation and guidance systems. The Third Academy has been conducting preliminary research into TERCOM (*dixing pipei zhidao*) since at least 1988, and, in 1991, published a thorough internal analysis of the Tomahawk's performance in the Gulf War.[15]

TERCOM capability is based on preprogramming the missile with digital maps. During flight, the missile's radar system matches the terrain surveyed to the computerized maps to ensure its course. This requires highly sophisticated digital mapping systems and powerful computers which will allow very accurate targeting. DSMAC uses a digitized satellite image of the target stored in the missile's computer system. In flight, the missile matches the stored image with images detected in flight. In general, the better resolution of the stored image, the more accurate the missile. China has already made great strides in development of digital mapping. To correct for en route navigation errors, Third Academy engineers are attempting to integrate GPS receivers with their inertial navigation units and TERCOM systems. Third

Academy engineers believe DSMAC can guarantee a CEP of less than 16 meters. Mid-course corrections through a combined GPS/INS system can further reduce the accuracy of cruise missiles to 10 meters.[16] There is some indication China is examining integrating combined GPS/GLONASS receivers on board their missiles as well.[17]

Other terminal sensor technologies under development include passive imaging infrared, CO_2 laser radar, millimeter wave, and synthetic aperture radar terminal sensors, as well as various composite systems. R&D into passive imaging infrared sensors is focused on matching a stored computerized image with a real infrared image detected by the missile.[18] Third Academy engineers have already laid the technical foundation for a CO_2 laser guidance system, to include the target recognition components.[19] Chinese aerospace engineers believe synthetic aperture radar, millimeter wave radar, infrared imaging, and laser radar guidance can result in an accuracy of one to three meters.[20]

There are other technologies the Chinese are researching to mask launch sites and enhance the ability of their missiles to penetrate defenses. Engineers are attempting to reduce the signature on cruise missile propellants, and conceal location of mobile land based strategic and tactical cruise missiles.[21] To further increase the penetration capability, the Third Academy is examining the technical prospects of integrating electronic and optical countermeasures onto cruise missiles in order to complicate adversarial detection, tracking, and discrimination systems.[22] The PLA Navy has conducted modeling of various anti-ship cruise missiles to evaluate performance against the SM-2 air defense missile system.[23]

The Third Academy is working with other defense industries to develop warheads which can effectively disable strategically important targets. Most significant is work on mastering submunition (*zidantou*) and guided dispensor (*zhidao busanqi*) technology. Submunitions can knock out aircraft in the open, crater runways to temporarily prevent aircraft from taking off and landing, or lay mines to make runway repair more difficult. They are also furthering R&D on more lethal unitary warheads for precision strikes against key facilities.[24] There are persistent signs that the Chinese intend to arm at least a portion of their cruise missile inventory with nuclear warheads. Engineers are also laudatory of cruise missiles equipped with high powered microwave (*dagonglu weibo*) warheads. If they are able to move beyond preliminary R&D on high powered microwave technology, such a directed energy warhead would be especially lethal against shipborne or airbased electronic systems.[25]

The Third Academy has upgraded its ability to design and manufacture highly complex cruise missiles. They are integrating the use of virtual reality (*xuni xianshi*) in cruise missile development, and are using increasingly sophisticated supercomputers to design the missiles. Third Academy manufacturing centers have imported some of the world's most advanced engineering workstations, and three, four, and five-axis computer numerically controlled machine tools.[26] CASC's world-class simulation facility in western Beijing also aids cruise missile development by theoretically reducing testing requirements by 40-60 percent and shortening overall development time by 30-40 percent.[27]

A new generation of low observable, extended range, faster, and more lethal missiles will probably be adapted for multiple platforms. Cruise missiles launched from airborne (H-6 or H-7) or naval platforms will result in significantly extended range. Perhaps more dangerous would be an ability to launch from a submerged submarine, which the Third Academy is closely examining.[28] It is believed that the first naval platform capable of submerged launching of cruise missiles will be the Song-class submarine, produced at the Wuhan shipyard.[29]

Extension of the range and target set of its cruise missiles will likely require a shift in organizational responsibilities. Conventional and nuclear air-launched land attack cruise missiles will most likely be subordinated to the PLA Air Force. Tactical and strategic attacks against sea platforms will continue to be a PLA Navy mission. The PLA Navy will expand its mission to include conventional and nuclear cruise missile strikes against land-based targets.[30] The greatest change in responsibilities may well be an expansion of the Second Artillery's mission to include strategic ground-launched land attack with cruise missiles equipped with both nuclear and conventional warheads. In support of this new mission, the Second Artillery's Fourth Research Institute has been modeling the ability of cruise missiles to penetrate air defense systems.[31]

The Third Academy approach is to develop variations of existing shorter-range airframes which encompass some advanced technologies. The Third Academy is believed to have advanced to model R&D and testing of some of these initial variants. First, the PLA has allegedly ordered the acceleration of a ground launched land attack version of the YJ-8A (C-802). This 135-kilometer range system would be the first to incorporate GPS-assisted navigation and TERCOM guidance. Some sources believe GPS aided navigation could result in cruise missiles like the YJ-8 to achieve accuracies of up to 10 meters. GPS-aided guidance could be augmented by terrain contour matching.[32]

Until they are able to field their next generation of cruise missiles, China may enter into an agreement with Russia for the purchase of antiship missiles. The sale of two Russian guided missile destroyers was discussed during Li Peng's visit to Moscow on December 28, 1996. The destroyers include the SS-N-22, designed to counter Aegis ships, and allegedly one cruise missile the U.S. Navy fears most because of its supersonic speed, maneuverability, and large warhead.[33] Russia's Raduga Design Bureau has also assisted the Third Academy in application of stealth technology to an unidentified air launched cruise missile.[34] Russia has also sent specialists to the Third Academy's 8359th Institute to overcome various technical obstacles in the development of new launch platforms.[35]

Ballistic Missile Research and Development.

In addition to cruise missiles, CASC is making rapid advances in increasing the survivability and lethality of ballistic missiles. China began its ballistic missile research and development program in the mid-1950s. Recently returned from the United States, Qian Xuesen submitted a proposal for missile development in 1956. The CMC approved the proposal but had to turn to the Soviet Union for technical assistance. Under the 1957 Sino-Soviet technical accord, the Soviet Union provided rudimentary missiles, associated launching equipment, and technical assistance. By 1960, however, the Sino-Soviet relationship deteriorated and all Soviet technicians returned to Moscow. Despite this setback, China test launched its first domestically produced missile, the DF-2A, in 1965. Over the course of the next 20 years, a series of liquid-fueled missiles were designed and tested to allow the PLA to strike U.S. bases in Japan (DF-2), the Philippines (DF-3), Guam (DF-4), and continental United States (DF-5). In the 1960s, China's missile industry began development of a solid-fueled sea launched ballistic missile, the JL-1. In the late 1970s, work began on a land-based version of the JL-1, the 1,800 kilometer range DF-21. Subsequent solid fueled missile development has focused on shorter range systems equipped with conventional payloads and longer range counterparts to the DF-4 and DF-5 ICBMs.[36]

China's preeminent organization for ballistic missile development is the CASC First Academy (*diyi yanjiuyuan*, or *yiyuan*). The First Academy, also known as the China Academy of Launch Technology (CALT), consists of an overall design and systems integration department, and 13 research institutes and seven factories which are responsible for engines, control technology, inertial systems, warheads, materials, testing, and launchers. With more than 27,000 personnel, the First Academy is the largest R&D organization within CASC. Leveraging its expertise in ballistic missiles, the First Academy, headquartered in the southern Beijing suburb of Nanyuan, has developed a range of satellite launch vehicles. In its work on solid systems, the First Academy depends upon the Fourth Academy in Hohhot, Inner Mongolia, for its solid motors. CALT is also supported by institutes and factories subordinated to various bases deep inside China. One of these bases, the Sanjiang Space Group in Hubei province, has developed its own complete ballistic missile system, the 300-kilometer DF-11 (also known as M-11).[37]

Today, the First Academy's research and development resources are devoted toward ensuring its nuclear ballistic missile force remains a viable deterrent in the face of emerging missile defenses. Since the initiation of work on the Strategic Defense Initiative in the late 1970s, Chinese missile engineers have been busily working to ensure their missile force retains its vitality.[38] In addition, the PLA, drawing profound lessons from the Gulf War, views conventional ballistic missiles as a crucial aspect of China's emerging deep attack (*zongshen*) capability.[39] To meet these objectives, the First Academy's research and development resources have focused on increasing the range of its solid fueled ballistic missiles, and increasing the survivability, accuracy, and lethality of its existing tactical, campaign or theater (*zhanyi*), and strategic short range missiles.[40]

China's program to modernize their nuclear intercontinental ballistic missile forces is well-documented. To briefly summarize, however, China's current ICBMs, able to strike targets throughout the United States, are liquid fueled. To offset vulnerabilities in their small ICBM force, the space and missile community is developing a new generation of mobile, solid fueled ICBMs. These include the DF-31 (8,000 kilometer range) and DF-41 (12,000 kilometer range) ICBMs, and the sea-based version of the DF-31, the JL-2. The nuclear tipped DF-31 will probably be deployed in experimental regiments around the turn of the century, and will have the ability to strike targets as far away as Guam and Hawaii. After successfully testing the DF-31, the space and missile industry will likely move the DF-41 to the next developmental stage.[41]

As CALT proceeds in its DF-31/DF-41 development, another top R&D priority is ensuring its current and future products are able to penetrate potential missile defenses and survive any counterattack. Probably the most well-known program is the development of multiple warhead technology. This includes multiple warheads against a single target, and multiple warheads against multiple targets with the option of equipping each with independent guidance sets. As of January 1996, CALT was in midst of developing multiple warheads, each with its own guidance system and maneuvering capability.[42]

R&D on multiple independent reentry vehicles (MIRVs) was initiated as early as 1970. Technical difficulties, however, have stalled the program. The First Academy renewed R&D in 1983, shortly after the SDI was announced in March 1983. The DF-5A, able to strike targets in the continental United States (CONUS), was the designated recipient of the MIRVs, although there is no evidence that they have been deployed. Critical to this effort is the miniaturization of warheads, a possible objective of tests at Lop Nur over the last few years. According to Chinese missile designers, real and decoy warheads (*jia dantou*) can be mixed using multiple warhead technology. Real warheads can be painted with radar absorbing materials that will weaken radar returns to maximize similarities between real and decoy warheads.[43]

CALT is developing maneuverable (*jidong biangui*) reentry vehicles to complicate missile defense tracking. While vehicles can maneuver at any time during flight, Chinese engineers see most utility in programming a reentry vehicle to maneuver in its terminal phase, 20-30 seconds before striking its target. A reentry vehicle has the ability to maneuver within a lateral range of 556-900 kilometers. Another maneuvering option discussed is to send the warhead up to a higher altitude after separation from the missile, slowly descending in a glide for a very long distance, and then finally diving toward the target. Chinese engineers are also addressing problems associated with maintaining accuracy after exo-atmospheric maneuvering.[44]

Another option Chinese designers have recently discussed is a fractional orbiting missile (*bufen guidao daodan*) in which a warhead is placed into a low earth orbit. When the time comes to attack a target, an engine can be fired to bring the warhead to bear against a target, creating great difficulties for an adversary's missile defense.[45] China conducted a feasibility study on a fractional orbital bombardment system (FOBS) in 1966. This system launches a missile into very low orbit, approximately 95 miles above earth. Before completion of first orbit, a retro-rocket retards the speed of the warhead, which hits the target with only a few minutes warning. In 1965, China had learned from an American report that the Union of Soviet Socialist Republics (USSR) was developing a forward operations base (FOB); this report was never confirmed. They explored launching a missile over Antarctica and penetrating the weakest point in the U.S. warning network.[46]

The Chinese have devoted resources toward countering potential deployment of air- and space-based high powered laser systems. First, Chinese engineers have carried out R&D into spinning their ballistic missiles to prevent concentration of a high powered laser on a single spot.[47] Another method under consideration as an explicit countermeasure to boost phase intercepters is fast burn booster (*suran zhutui*) technology for China's next generation of solid fueled strategic ballistic missiles. Chinese engineers, however, caution designers about potential quality control problems related to stage separation and accuracy, and suggest this technology should be divided into three stages based on the pace of foreign missile defense developments.[48]

Missile analysts view depressed trajectories (*yadi guidao*) as another option to counter space-based and exoatmospheric upper-tier missile defense systems. Intercontinental ballistic missiles (ICBMs) often reach altitudes of 2,000 kilometers on a normal trajectory. However, launching a missile at a depressed trajectory could allow the missile to achieve only a 100 kilometer altitude which complicates some space-based systems' ability to penetrate the atmosphere. Testing and modeling has been done on the DF-3, which normally has a range of 2,780 kilometers, with a maximum altitude of 550 kilometers. With depressed trajectory, the DF-3 travels 1,550 kilometers at 100 kilometers altitude.[49]

Jammers can also increase the ability of missiles to penetrate missile defenses and strike their targets. Passive jamming includes the use of chaff (*jinshu botiao*) to confuse enemy sensors, and tests have shown that 122 kilograms of chaff can affect an area 320 kilometers wide and 720 kilometers long. Other measures under investigation include electronic and infrared countermeasures on board reentry vehicles, as well as carrying out hard kills against enemy theater missile defense (TMD) radars through the use of anti-radiation missiles.[50]

Finally, the Second Artillery and CASC have conducted modeling exercises and simulations to test China's ability to break though the wide range of projected U.S. TMD deployments. This includes developing a capability to attack airborne missile defense platforms as well. Modeling has focused on using combinations of surface-to-surface, air-to-surface, and sea-to-surface missile systems.[51] After computer simulations and modeling exercises, CALT is confident that its maneuverable theater ballistic missile re-entry vehicles can defeat opposing PAC-2+ systems.[52]

Protection against attack, or survivability, is another important aspect of ballistic missile modernization. Survivability measures include silo hardening, mobility (road/rail); scattered and camouflaged launch sites.[53] Chinese R&D into camouflage technology is explicitly intended to counter U.S. air and space-based reconnaissance platforms.[54] CALT officials have evaluated various deployment and launching modes for their strategic missile survivability, and many have concluded China's future interests can best be served through a mix of silo-based and mobile launchers, with an emphasis on rail-launched systems. Also emphasized was integration of decoys and concealment technology in their missile development and deployment.[55]

Another vital aspect of ballistic missile modernization is accuracy. The First Academy is striving to increase the precision of their ballistic missiles through a variety of means. First, the First Academy is developing better gyros and inertial measurement units. With assistance from Russia, CALT is working on applying beryllium technology to their inertial measurement units to decrease the weight of their guidance sets.[56] Reducing the weight of attitude control and guidance sets is crucial for development of multiple warheads on a single missile, with each reentry vehicle equipped with its own guidance system.

Like the Third Academy, CALT is exploiting the U.S. NAVSTAR Global Positioning System (GPS) constellation of satellites. With a receiver installed onto the missile, these GPS satellites provide en route updates to the control system to correct trajectory errors. Some estimates indicate GPS assistance can help achieve an accuracy of 100 meters or better.[57] At least two tests of an on-board GPS trajectory reference system had been conducted as of 1995.[58] According to one authoritative estimate, GPS-aiding of ballistic missiles can improve overall ballistic missile accuracy by 20-25 percent.[59] Engineers have also developed an integrated NAVSTAR GPS and GLONASS receiver.[60] If mounted on a missile, the benefits of such a receiver are great. Should an adversary jam or turn off GPS, the missile system can still rely on the GLONASS updates. The First Academy is integrating GPS onto their mobile launchers to further enhance the initial reference point and increase accuracy.[61] Engineers have also carried out applied research into applications of differential GPS (*chafen quanqiu dingwei xitong*) as another measure to further increase the accuracy of ballistic missiles.[62]

To improve accuracy even further, the space and missile industry is carrying out preliminary R&D on digital scene matching terminal guidance (*jingxiang quyue xiangguanqi*), which uses on-board computers to correlate stored images with landmarks. First Academy engineers believe digital scene matching can result in a 5-12 meter accuracy for their ballistic missiles.[63] The Second Artillery has expressed support in scene matching terminal guidance, which would require the manueverable reentry vehicle previously discussed.[64]

The First Academy, in conjunction with COSTIND and the Second Artillery, is examining a wide range of warheads in order to make its missiles more lethal and able to exploit weaknesses in foreign TMD systems now under development. The First Academy's warhead design institute (13th Research Institute, or Beijing Institute of Special Electromechanics) is carrying design work on runway penetrators to support offensive counterair missions.[65] Indications exist that the 13th Research Institute is working to overcome problems associated with submunition dispensing warheads, which can cause damage over a significantly greater area than a unitary warhead. Other warheads under development reportedly include mine-laying payloads, an electromagnetic pulse warhead (i.e., HPM) to take out enemy C^4I equipment, and penetration warheads to target fortifications deep underground. Some reporting indicates the 600 kilometer range DF-15 could feasibly achieve an accuracy of 30-45 meters.[66]

Second Artillery.

The PLA's primary arm for strategic attack is the Second Artillery (*erpao*). Its traditional realm has been in nuclear strategic missiles. However, since the conclusion of the Gulf War, Chinese planners have seen value in expanding the 2nd Artillery's mission to include conventional strikes against high value strategic targets. The Second Artillery, with an estimated 90,000 personnel, consists of headquarters elements, six launch bases, four research institutes, two command academies, and possibly an early warning unit.[67] The headquarters is located in Qinghe, a suburb to the north of Beijing. Commanded by LTG Yang Guoliang, the Second Artillery headquarters complex consists of headquarters, political, logistics, and equipment technology departments, each headed by a PLA major general. The headquarters complex maintains contact with subordinate units through its own communications regiment. Three deputy commanders assist Lieutenant General Yang Guoliang in his responsibilities. The Second Artillery maintains liaison offices within the China Academy of Engineering Physics, CASC's First Academy's liquid fueled logistic vehicle (LV) integration factory (211 Factory) and CASC's 307 Factory (solid fueled system integration).[68]

The Second Artillery headquarters complex oversees six launch bases which are division-sized elements, each commanded by a major general. A base headquarters structure mirrors that of the national headquarters in Qinghe with Headquarters, Political, Logistics, and Equipment and Technology Departments (Bu). Base-level equipment and technology departments (*zhuangji bu*) oversee launch site management (*zhendi guanli*); warhead stations (*dantizhan*); and launch repair facilities.

Each base has two or three subordinate brigades (*lu*) and regiment-level special departments (*chu*) responsible for chemical defense, communications, training, security, and weather. Each brigade, commanded by a colonel, has up to four launch battalions (*fashe ying*). Missile brigades are generally structured by type of missiles. In other words, one brigade only has one type of missile, thus facilitating maintenance and specialization. There are least 13 brigades in the Second Artillery. In addition, each base has training (*jiaodaodui*) and nuclear warhead (euphemistically referred to special equipment) maintenance units. Each base is directly subordinate to the Second Artillery commander in Qinghe, although they do receive support from the military regions. The total number of nuclear warheads maintained by the Second Artillery is unknown.[69]

80301 Unit. The 80301 Unit is headquartered in Shenyang, Liaoning province, and consists of three launch brigades, each with up to three launch battalions. Its DF-21 and DF-3 IRBMs cover the Korean peninsula and Japan, to include Okinawa. At least one of the brigades, equipped with 1,800 kilometer range DF-21s (CSS-5), is concentrated in the area of Tonghua, approximately 80 kilometers north of the North Korean border. R&D on the DF-21 began in 1967 and had its first successful test in 1985. Shortly thereafter, the DF-21 was deployed into an experimental regiment. The DF-21 follow-on, the DF-21A, is expected to have a 1800 kilometer range, with a 600 kilogram warhead.[70]

80302 Unit. The 80302 Unit, also known as 52 Base, is headquartered in the mountain resort town of Huangshan, Jiangxi province, and is the Second Artillery's most important unit for conventional long-range precision strikes against Taiwan. The Huangshan base has at least two, but probably three brigades. The 52 Base's most well-known brigade, the 815th, garrisoned in Leping, Jiangxi province, is equipped with the newest addition to the Second Artillery's inventory, the DF-15. The 600-kilometer range DF-15s gained notoriety during the March 1996 missile exercises off the coast of

Taiwan. During a wartime situation, the 815th Brigade's missiles scatter to pre-surveyed launch sites (*zhendi*) in Fujian province in order to range the entire island of Taiwan. China's total DF-15 inventory is estimated as being from one hundred to several hundred missiles. Rail is the usual way of moving missiles from garrison to field deployments.[71]

80303 Unit. The 80303 Unit, headquartered in the eastern suburbs of Kunming, Yunnan province, has two brigades. One of the brigades, equipped with DF-21s, is located in Chuxiong, approximately 100 kilometers west of Kunming, while the other is located at Jianshui, south of Kunming. These missiles are in range of several targets in India and Southeast Asia.[72]

80304 Unit. The 80304 Unit, headquartered in Luoyang, Henan province, has three brigades. At least one brigade, located in the area of Luoning, is equipped with the 12,000 kilometer DF-5 ICBM, able to strike targets throughout the United States and Europe. One other brigade is located near Sundian.[73]

80305 Unit. Headquartered in the remote western Hunan provincial city of Huaihua, the 80305 Unit is the second command equipped with ICBMs. Its two subordinate brigades are located south of Huaihua, near the town of Tongdao.[74]

80306 Unit. The 80306 Unit, headquartered in Xining, Qinghai province, has three brigades, including one potential experimental regiment, concentrated in the Datong, Delingha, and Da Qaidam areas. Indications exist that the 80306 Unit may upgrade to the DF-21. The 80306 Unit is able to target sites in the former Soviet Union and India.[75]

Institutes and Colleges. The Second Artillery also has one engineering design academy and four research institutes to solve problems associated with operations, transporter erector launchers (TELs), and logistics (First Institute), command automation, targeting, and mapping (Third Institute), and missile and warhead engineering design. The Second Artillery's Command College in Wuhan prepares officers for leadership positions within headquarters elements and launch brigades. The Engineering College in Xian educates technicians largely associated with equipment and technology departments at various headquarters and field units. The Second Artillery's Academy of Engineering Design works closely with CASC in evaluating overall design work.[76]

Doctrine. The 2nd Artillery has separate doctrines for its nuclear and conventional missile forces. First, ballistic missiles are the key to providing strategic nuclear deterrence. In the past, the Second Artillery is thought to have followed a doctrine which required only a small number of warheads necessary to cause unacceptable damage on a handful of enemy cities. This minimal deterrence doctrine, however, appears to be evolving toward a limited nuclear deterrence (*youxian weishe*) approach. Limited nuclear deterrence differs from minimal deterrence in that the latter relies on simple countervalue strikes. Very few warheads are necessary. Limited deterrence doctrine has counterforce and warfighting elements with intended targets including enemy missile bases, C^4I centers, and strategic warning assets. Limited deterrence requires a greater number of smaller, more accurate, survivable, and penetrable ICBMs and sea-launched ballistic missiles (SLBMs); a ballistic missile defense capability to protect its limited deterrent force; and ASAT assets to hit enemy satellites. According to nuclear weapon strategists, a space-based early warning capability is needed to speed up reaction time.[77] China's strategic modernization R&D support this shift toward a limited warfighting approach to nuclear warfare.

The Second Artillery's conventional mission currently provides China its only reliable means of long-range attack against an enemy's high value targets. A growing sector of the PLA believes ballistic missile strikes against strategic targets are the best way to even out the playing field when fighting against a technologically superior force. Increasingly accurate ballistic missiles, aerospace engineers assert, can be targeted against enemy air defense nodes and even against aircraft carriers. Chinese analysts view a large arsenal of ballistic missiles as a relatively cheap means to saturate an expensive theater missile defense architecture.[78] According to one estimate, the Second Artillery's missile supplier, CASC, will have the capacity to produce 1,000 new ballistic missiles within the next decade.[79]

Over the horizon missile strikes against critical nodes can paralyze an enemy's combat capacity. Strikes against important targets also drive up the human and material costs and put psychological pressure on enemy troops and domestic constituency. Targets in a campaign include C^3I nodes, weapon control centers, high value air assets on the ground, logistics bases, and important sea combat platforms. Strikes against air bases can enhance the PLAAF's ability to gain air superiority by temporarily degrading an enemy's ability to generate sorties. Other important targets include logistics bases.[80]

Missile strikes would likely be conducted as soon as the Chinese leadership believes war is inevitable. Absorbing lessons learned from the Gulf War, the PLA believes an enemy such as the United States is most vulnerable when it is deploying forces and logistics to the area of operations. A preemptive strike during this phase, many PLA strategists believe, will significantly offset the an enemy's technological advantages.[81]

The Second Artillery clearly understands the implications of counterstrikes after missile attacks. Chinese engineers are devising means to reduce the time a missile launcher is exposed at a pre-surveyed launch site. In one concept, adjustments are made to the TEL which allows for the missile erection within one minute and retraction within a minute. Reductions in TEL weight will increase mobility and allow transport over lower classes of roads and bridges.[82] To increase accuracy of their missiles, TELs are being equipped with GPS receivers.[83]

ENDNOTES - CHAPTER 4

1. "China Replacing Nuclear Warheads on Some Missiles," *Jane's Defense Weekly*, January 27, 1994.

2. See Appendix I on CASC organization for details on Third Academy.

3. Yu Yongbo, ed., *China Today: Defense Science and Technology*, Vol. 2, Beijing: National Defense University Press, 1993, pp. 508-541.

4. Wang Jianmin and Zhang Zuocheng, "Jiasu Jibenxing Xiliehua Jincheng Nuli Fazhan Woguo Feihang Daodan Shiye" ("Rapid Progress in Series Development of China's Cruise Missile Industry"), *Zhongguo Hangtian*, September 1996, pp. 12-17.

5. Xu Ande, "Significance and Economic Benefits of Developing a Modern Cruise Missile for China," *Xiandaihua* (Modernization), Vol 14, No. 4, April 1992, in *JPRS-CST*-92-013.

6. Wang Jianmin and Zhang Zuocheng, pp. 12-17.

7. See Zheng Wanqian, "Xunhang Daodan Wuqi Xitong de Guanjian Jishu" ("Key Technologies of Cruise Missile Weapon Systems"), *Zhongguo Hangtian*, July 1996, pp. 42-45.

8. Yu Wenman and Liu Chunbao, "Sea Target Characteristic Study," *Hangtian Qingbao Yanjiu Baogao Xilie Wenzhai*, October 1987, pp. 252-259.

9. Si Xicai, "Research on Long Range Antiradiation Missile Passive Radar Seeker Technology," *Zhanshu Daodan Jishu (Tactical Missile Technology)*, 1995, Vol. 2, pp. 42-52, in *CAMA*, 1995, Vol. 2, No. 5. Other studies on specific approaches to ARM technology include Yang Huayuan, "Study on Superwideband High Accuracy Microwave DF System," *Daojian yu Zhidao Xuebao*, February 1995, pp. 7-12, in *CAMA*, Vol. 2, No. 5. There are also strong indications that SAST's system engineering organization, the Shanghai Institute of Electro-Mechanical Engineering, is carrying out preliminary R&D on a long-range air-to-air antiradiation missile for targeting airborne early warning platforms, such as the U.S. E-3 AWACS or Taiwan's E-2C's. Engineers note critical technologies for development of a long-range ARM include a passive seeker with a sensitivity of greater than 100dB, as well as monolithic microwave (*danpian weibo*), gallium arsenide, very large scale, and very high speed integrated circuits (MMIC, GAAS, VLSIC, VHSIC'S). The seeker makes up for greater than 50 percent of the R&D and production costs for an ARM. At least one Second Academy entity which has conducted work on antiradiation missile seeker technology is the Beijing Institute of Remote Sensing Equipment (probably the CASC 25th Research Institute).

10. China's nuclear-tipped cruise missile R&D is discussed in 1993 DoD report which warns too much of SDIO's effort is focused on countering ballistic missiles, ignoring an increasing cruise missile

threat. See "Cruise Missiles Becoming Top Proliferation Threat," *Aviation Week & Space Technology*, February 1, 1993, pp. 26-27.

11. For a good overview of the Third Academy's direction in cruise missile development, see Zheng Wanqian, "Xunhang Daodan Wuqi Xitong de Guanjian Jishu" ("Key Technologies of Cruise Missile Weapon Systems"), *Zhongguo Hangtian*, July 1996, pp. 42-45.

12. 3rd Academy's Zhu Yubao, "Study of LRCS Contour Design for Conventional Cruise Missiles, *Zhanshu Daodan Jishu (Tactical Missile Technology)*, Vol. 3, 1994, pp. 1-11; and Huan Shengguo, "Application of Stealth Technology in Missiles," *Shanghai Hangtian*, June 1996, pp. 57-59, in *FBIS-CST*-97-002.

13. Long Yuzhen, "Characteristics and Design Schedule of Scramjet Engine For Hypersonic Cruise Missile," *Feihang Daodan*, August 1997, pp. 29-37, in *CAMA*, 1997, Vol. 4, No. 6; Shang Aidong, "A Supersonic Combustion Ramjet Missile," *Feihang Daodan*, June 1996, pp. 40-48, in *CAMA*, Vol. 1, No. 5; Dong Yuejuan, "Technical Path for Developing A Scramjet Engine Propulsion System," *Feihang Daodan*, Vol. 1, 1996, pp. 42-50; Si Tuming, "Performance Analysis of Waverider Hypersonic Cruise Missile With Integrated Scramjet Engine," in *Zhanshu Daodan Jishu (Tactical Missile Technology)*, 1996 (1) pp. 45-50; and Si Tuming, "Supersonic Cruise Missile and Ramjet Engines," in *Feihang Daodan*, Vol 1, 1996, pp. 35-41. Long Yuzhen and Si Tuming are from the 31st Institute (Beijing Institute of Power Machinery) of the Third Academy, the primary entity responsible for R&D of cruise missile propulsion systems. *Feihang Daodan* is the Third Academy's most well-known technical journal and China's most authoritative voice for cruise missile development. Also see Huang Zhicheng and Qiu Qianghua, "Prospects for Hypersonic Flight," *Liuti Lixue Shiyan Yu Celiang (Experiments and Measurements in Fluid Mechanics)*, 1997, Vol. 11, No. 1, pp. 6-11, in *CAMA*, Vol. 4, No. 4; and Huang Zhicheng, Qiu Qianghua, and Yuan Shengxue, "Background, Recent Status, and Missions of Supersonic Combustion Research," paper presented at the November 1996 National Aerospace Aerodynamics Society Meeting, in *CAMA*, 1997, Vol. 4, No. 5. Technical writings indicate COSTIND's Beijing Institute of Systems Engineering, specifically Huang Zhicheng and Qiu Qianghua, also plays a major role in hypersonic vehicle R&D, especially as part of China's aerospace plane program.

14. If the Chinese are looking to develop a 1,500 kilometer missile, the Russian 1,500 kilometer range AS-15 could be used as a model. Some modifications would have to made to enable it to launch from the ground. The Tomahawk has a 450 kilometer range, while the U.S. AGM-86B has a 3,000 kilometer range.

15. Zhong Longyi, "Zuhe Daohang Xitong he Bingxing Duoji Xitong Zai Xunhang Daodanzhong de Yingyong," ("Application of Combined Navigation Systems on Cruise Missiles") in *Hangtian Qingbao Yanjiu (China Information Research)*, 1993 (3), pp. 432-445. Zhong is from the Third Academy's 8357 Research Institute, responsible for cruise missile control systems. For an example of research on runway penetrators, see Yang Bingwei, "Structural Design Problems and Test Methods of Anti-Runway Penetrators," *Hangtian Keji Qingbao Yanjiu Baogao*, Vol. 5, May 1995, pp. 288-303, in *CAMA*, Vol. 3, No. 6.

16. Zheng Wanqian, p. 43. TERCOM requires highly sophisticated digital mapping systems and powerful computers. COSTIND and the Second Artillery have made significant achievements in both areas. See Wang Yongming, "Introduction to Military Electronic Maps," *Xiaoxing Weixing Jisuanji Xitong*, (Mini-Micro Computer Systems), August 1995, pp. 12-18, in *FBIS-CST-96-001*. Wuhan Technical University of Surveying and Mapping is one institute involved in digital mapping. Also see Jing Shaoguang, "GPS/SINS Integrated Navigation System for Cruise Missiles," *Xibei Gongye Daxue Xuebao*, 1997, Vol. 15 (1), pp. 79-83, in *CAMA*, 1997, Vol. 4, No. 6.

17. Guan Dexin, "The Investigation of Compatible Receiver For GPS and GLONASS," *Xitong Gongcheng Yu Dianzi Jishu*, 1996, Vol. 18, No. 7, pp. 69-74, in *CAMA*, Vol. 3, No. 6; and Sheng Jie, "Demonstration of Navigation Performance of GLONASS/GPS Composite Receivers," *Weixing Yingyong*, February 1994, pp. 56-59, in *CAMA*, 1994, Vol. 1, No. 5.

18. See Wang Jianmin, "Work Hard to Develop Cruise Missile Industry," *Zhongguo Hangtian*, September 1996, pp. 12-17; Sun Qingguang, "Study on Laser Imaging Guidance," *Feihang Daodan*, March 1995, pp. 46-50, in *CAMA*, Vol. 2, No. 3; Liu Yongchang, "Infrared Imaging Precision Seeker Technology," *Hongwai Yu Jiguang Jishu (Infrared and Laser Technology)*, 1996, Vol. 25, No. 3, pp. 47-53, in *CAMA*, Vol. 3, No. 6; Zhao Jun, "Applied Research Into Laser Imaging Guidance Technology Development," *Hangtian Qingbao Yanjiu*, HQ-96039, in *CAMA*, 1997, Vol. 4, No. 2; and Li Jin, "Development of Infrared Focal Plane Array Imaging Technology," in *Feihang Daodan Qingbao Yanjiu Baogao Wenzhai (Cruise Missile Information Research Reports)*, December 1996, pp. 190-209, in *CAMA*, Vol. 4, No. 6. Leading the infrared imaging effort is the Third Academy's Tianjin Jinhang Technical Physics Institute.

19. Sun Qingguang, "Jiguang Chengxiang Zhidao ji Ganrao Moushi de Yanjiu" ("Research Into Laser Imaging Guidance and Jamming"), in *Hangtian Qingbao Yanjiu*, HQ-93017, pp. 228-241.

20. Zheng Wanqian, p. 43.

21. Zhang Haixiong, "ADN: Oxidizer for A Low Signature Propellant," in *Feihang Daodan*, July 1996, pp. 35-38, in *CAMA*, 1996, Vol. 3, No. 6; and Lu Xiaohong, "Camouflage and Concealment Technology of Mobile Missile Launchers and Ground Equipment," *Harbin Institute of Technology Journal*, December 1996, pp. 266-277, in *CAMA*, Vol. 4. Lu is from the Third Academy's Beijing Institute of Special Machinery, responsible for cruise missile launchers.

22. Zhou Yunwu, "Jamming Techniques for Cruise Missile Penetration," *New Electronic Warfare Technology and Intelligence Reform Abstracts*, 1995 (10), pp. 143-147, in *CAMA* 1996, Vol 3, No. 6. Zhou is from the Third Academy's Beijing Institute of Electro-mechanical Engineering.

23. Xu Cheng, "Research on Optimal Penetration Model of A Chinese Anti-ship Missile," paper presented at November 1997 conference of National Missile Designers Information Network Conference, in *CAMA*, Vol. 5, No. 3.

24. Dong Yuejuan, "Zhidao Busanqi Jiqi Yingyong Fenxi" ("Applied Analysis of Guided Dispensors"), *Hangtian Qingbao Yanjiu*, HQ-93016, pp. 214-227.

25. High-powered microwave weapons are discussed in more detail in Chapter 5 of this study. See Zheng Wanqian, p. 43.

26. Tian Baolong and Li Wengang, "Feihang Daodan CAM Chejian Danyuan Xitong" ("Cruise Missile CAM Workshop Unit System"), *Zhongguo Hangtian*, April 1993, pp. 44-46; Xu Haijiang, "Virtual Reality and Its Application in Development of Cruise Missiles," in *Feihang Daodan*, 1996 (8), pp. 1-9; Wang Zhenhua, "Parallel Computation on Supercomputers for Axisymmetric Interaction Flow," *Yuhang Xuebao (Journal of Astronautics)*, January 1995, pp. 43-45, in *JPRS-CST*-95-005.

27. Li Weiliang, "Jiang Zemin dao Beijing Fangzhen Zhongxin Zhouyan" ("Jiang Zemin Inspects Beijing Simulation Center"), *Zhongguo Hangtian Bao*, January 17, 1994, p.1; Li Li, "Chinese Simulation Technology Among Leaders Worldwide," *Liaowang Zhoukan*, August 16, 1993, pp. 4-5, in *JPRS-CST*-93-017. American aerospace representatives who have been allowed access have remarked that the CASC Beijing Simulation Center is very close in capabilities to Western simulation facilities.

28. Zheng Wanqian, p. 44; Zhang Lingxiang, "Cruise Missile Container Launching Technology," *Hangtian Fashe Jishu (Space Launch Technology)*, January 1996, pp. 1-8, *CAMA,* 1996, Vol 3, No. 6. Zhang is from the Beijing Institute of Special Machinery (Beijing Tezhong Jixie Yanjiusuo).

29. *Jane's Navy International*, April 1, 1997, p. 14.

30. Liu Kejun, "Information Warfare Challenge Faced By Navy," *Zhongguo Dianzi Bao*, October 24, 1997, p. 8, in *FBIS-CHI*-98-012.

31. One Taiwan source explicitly asserts land attack cruise missiles will be assigned to the Second Artillery. See "Mainland Acquisition of Russian Weapons Viewed," *Lien-Ho Pao*, April 29, 1996, in *FBIS-CHI-96-086*; also see Sun Xiangdong and Qin Xiaobo, "Operational Efficiency Analysis of SAMs Against Cruise Missiles," *Xitong Gongcheng Yu Dianzi Jishu (Systems Engineering and Electronics)*, October 1996, pp. 59-63, in *FBIS-CST*-97-013.

32. Jason Glashow and Theresa Hitchens, "China Speeds Development of Missile With Taiwan Range," *Defense News*, March 4-10, 1996, p. 1.

33. Bill Gertz, "U.S. Not Against Russia-China Deal," *Washington Times*, January 11, 1997, p. A4.

34. "Russian Missile Assistance to China," *Flight International*, August 31, 1995. As a side note, Raduga is also working on hypersonic standoff cruise missiles.

35. "BasanWuJiu Suo Yu E'luosi Zhuanjia Jinxing Jishu Jiaoliu" ("The 8359 Institute and Russian Experts Carry Out Technical Exchanges"), *Zhongguo Hangtian Bao*, July 7, 1994, p. 2. The 8359th, also known as the Beijing Institute of Special Machinery, is responsible for development of cruise missile launching equipment, to include air, ship, submerged submachine, and ground equipment.

36. For a complete history, see *China Today: Defense Science and Technology*, pp. 256-311; *China Today: Space Industry*, pp. 90-116; Lewis and Xue; and Lewis and Hua, pp. 5-40.

37. See Appendix I for a complete organizational breakout. A large number of Sanjiang's M-11s are thought by many to have been sold to Pakistan in the early 1990s.

38. For an example of a study done in response to the U.S. Brilliant Pebbles project, see Chen Jianxiang, "Threat Area of Brilliant Pebbles

and Attacking ICBMs," *Xitong Gongcheng Yu Dianzi Jishu*, 1994, Vol. 16, No. 8, pp. 69-80, in *CAMA*, 1994, Vol. 1, No. 5.

39. Liu and Yang, pp. 4-26; and Wang Jixiang, "Inspiration for Chinese Ballistic Missile Development From the Gulf War," *Hangtian Keji Qingbao Yanjiu Baogao Xilie Wenzhai*, April 1994, pp. 49-56, in *CAMA*, Vol. 3, No. 6.

40. CALT defines tactical missiles as having a range of 10-300 kilometers (i.e., the DF-11), campaign missiles as having a range of 300-1,000 kilometers (i.e., the DF-15), and strategic short-range missiles as having a range of 1,000-2,000 kilometers (i.e., the DF-21). See Gan and Liu, p. 30.

41. The best account of China's ballistic missile modernization program is Lewis and Hua.

42. Gan and Liu, p. 42.

43. *Ibid.*, p. 46; Gui Yongfeng, "Penetration of Tactical Ballistic Missile's Decoy," *Hubei Hangtian Keji (Aerospace Hubei)*, February 1994, pp. 36-38, in *CAMA*, 1995, Vol. 2, No. 1; and Li Hong, "Motion Characteristics of Atmospheric Reentry Ballistic Missile Warheads and Their Applications To Heavy Decoy Design," *Jiangnan Hangtian Keji (Jiangnan Space Technology)*, 1997 (1), pp. 26-30, in *CAMA*, 1997, Vol. 4, No. 3.

44. Gan and Liu, p. 43. Also see Cai Yuanli, "Research on Trajectory Recovery in Exo-Atmospheric Flight," *Daodan Yu Hangtian Yunzai Jishu (Missiles and Space Vehicles)*, March 1995, pp. 10-15, in *CAMA*, Vol. 2, No. 5; and Zhao Hanyuan, "Simulation, Analysis of Maneuverable Reentry Vehicles," *Yuhang Xuebao*, January 1, 1997, pp. 96-99, in *FBIS-CST*-97-012. Zhao is from the National University of Defense Technology.

45. Gan and Liu, p. 44.

46. Lewis and Hua, p. 17.

47. Meng Daikui, "Simulation of Control and Guidance of Spinning Missiles," *Xitong Gongcheng yu Dianzi Jishu*, 1994, Vol. 5, No. 3, summarized in *CAMA*, 1995, Vol. 2, No. 1; Wan Chunxiong and Yang Xiaolong, "Identification of Flight Disturbances on Spinning Missiles," *Zhanshu Daodan Jishu (Tactical Missile Technology)*, March 1995, pp. 1-8, in *CAMA*, Vol. 2, No. 3. For a general assessment on methodologies to protect missile systems against high-powered lasers, see Ji Shifan, "Protection of Missiles Against Lasers," *Daodan yu Hangtian Yunzai*

Jishu, 1996 (5), pp. 35-42, in *CAMA*, Vol. 4, No. 1. Ji's research concentrated on the effects of high-owered lasers on a variety of materials and opto-electronic systems.

48. Wang Jixiang, "Fast Burn Boost Strategic Ballistic Missile Technology," *Aerospace S&T Intelligence Studies Abstracts* (2), Vol. 92 (4), pp. 68-78, in *CAMA*-96, Vol 3, No. 6. Wang is from the Beijing Institute of Space Systems Engineering (Beijing Yuhang Xitong Gongcheng Yanjiusuo); also see Qin Guangming, "Application of Slotted Tubular Grain in Fast Burn Solid Motors," *Bingong Xuebao (Ordnance Journal)*, 1996, Vol. 18, No.2, pp. 41-43, in *CAMA*, Vol. 3, No. 6. Qin is from the Xian Institute of Modern Chemistry.

49. Du Xiangwan, "Ballistic Missile Defense and Space Weapons," in *Quanguo Gaojishu Zhongdian Tushu, Jiguang Jishu Linghuo, (National High Technology Key Reference— Laser Technology Realm)*, Beijing: National Defense University Press, undated, p. 285.

50. Gan and Liu, p. 45. Also see Zhang Demin, "Study on Penetration Techniques on New Generation Ballistic Missiles," *Xinjunshi Gemingzhong Daodan Wuqi Fazhan Qianjing*, November 1996, pp. 18-24, in *CAMA*, Vol. 4, No. 2.

51. Jin Weixin, "Mathematical Modeling of Tactical Surface to Surface Missiles Against TMD," *Systems Engineering and Electronic Technology*, 1995, Vol. 17 (3), pp. 63-68, in *CAMA* 1995, Vol 2, No.3.

52. Zhang Demin and Hou Shiming, ""Simulation Research of Offensive and Defensive Capability of Conventional Manuevering Reentry Missile," *Xitong Gongcheng Yu Dianzi Jishu*, 1997, Vol. 19 (4), pp. 45-49, in *CAMA*, 1997, Vol. 4, No. 5. Full translation in *FBIS-CHI*-97-272. Zhang is from the Beijing Electromechanical Engineering Design Department, also known as the CASC Fourth Systems Design Department. According to one evaluation, PAC-2 has a probability of kill of 10-25 percent against an unidentified tactical ballistic missile. See Zhao Yuping, "Probability of PAC-2 Intercepting a Certain Tactical Ballistic Missile," paper presented at November 1997 conference of National Missile Designers Specialist Network, in *CAMA*, Vol. 5, No. 3.

53. Gan and Liu, pp. 42-45.

54. Lu Xiaohong, "Camouflage and Concealment Technology for Mobile Launchers and Ground Equipment of Strategic and Tactical Missiles," *Aerospace Industry Press*, HQ-96034, 1996, in *CAMA*, Vol. 4, No. 2. The key institute for CCD technology related to missile launchers

is the Beijing Institute of Special Machinery (CALT 15th Research Institute).

55. Wen Longzhi, "Evaluation of the Strategic Missile Survivability," *Aerospace Science Intelligence Studies Report Abstracts*, No. 5, 1995, pp. 353-368, in *CAMA*, Vol. 3, No. 6.

56. See for example, Chen Zhaoxi, "Manufacturing Technology of Beryllium Inertial Measuring Units," *Hangtian Gongyi*, March 1993, pp. 39-41, in *CAMA*, Vol. 1, No. 5. Chen is from the Beijing Institute of Control Instruments (Beijing Kongzhi Yibiao Yanjiusuo). The United States has used beryllium in its strategic missile guidance sets since the 1960s.

57. Wang Chiung-hua, "The Military of Communist China Plans to Use Global Positioning System to Improve Precision of Missiles," *Chung-Yang Jih-Pao*, December 24, 1997, p. 10, in *FBIS-CHI-97-361*.

58. For a summary of test results, see Sun Mei, "GPS For Evaluating Inertial Measurement Unit Errors," *Hangtian Congzhi (Aerospace Control)*, 1995, Vol. 13, No. 1, pp. 69-75, summarized in *CAMA*, 1995, Vol. 2, No. 4. Also see Wang Shuren, "Principles of Onboard GPS Navigation Transponders," *Hangkong Dianzi Jishu*, undated, pp. 20-23, in *CAMA*, 1994, Vol. 1, No. 5. Wang is from the Second Artillery's Academy of Engineering.

59. Scott Pace, Gerald Frost, Irving Lachow, David Frelinger, Donna Fossum, Donald Wassem, and Monica Pinto, *The Global Positioning System: Assessing National Policies*, RAND: Critical Technologies Institute, p. 68.

60 "GNSS=GPS+GLONASS," *Zhongguo Hangtian Bao*, November 6, 1996, p. 4. The receiver exploits a total of 48 GPS and GLONASS satellites to achieve a 0.5 CEP for differential positioning accuracy or a seven meter CEP for no differential. Differential positioning system includes a reference station within 100-200 kilometers of the launch site. Also see Sheng Jie, "Performance Demonstration of Combined GLONASS/GPS Receivers," *Weixing Yingyong (Satellite Applications)*, February 1994, pp. 56-59, in *CAMA*, Vol. 1, No. 5. There is no hard evidence China has integrated these receivers on a missile system, or that China has a differential positioning station within 200 kilometers of the missile launch site. However, there are indications China uses a differential GPS system in support of its space launch program.

61. Zhang Hu, "Application of GPS in Missile Maneuvering Positioning," *Zhongguo Yuhang Xuehui Fashe Gongcheng Yu Dimian*

Shebei Wenzhai (*China Astronautics Society Launch Engineering and Ground Equipment Abstracts*), November 1993, in *CAMA*, Vol. 1, No. 5.

62. Li Yonghong, "Ballistic Trajectory Determination Using the Differential Global Positioning System," *Binggong Xuebao*, 1997, Vol. 18 (4), pp. 372-374, in *CAMA*, Vol. 5, No. 2.

63. It is not clear how far engineers have gone in their preliminary research in this type of ballistic missile terminal guidance. See Gan and Liu, pp. 68-69. TERCOM is only effective over land where landmarks are used as reference points.

64. Wang Honglei, "Optical Image Guidance Technology," *Zhidao yu Yinxin*, January 1995, pp. 34-37, in *CAMA*. Vol. 2, No. 3. Wang is from the Second Artillery. There has been some Western open source reporting on China's research into optical terminal guidance for its missiles. Russia's SS-X-26 utilizes a similar system.

65. Yang Bingwei, "Structural Design Problems and Test Methods of Anti-Runway Penetrators," *Aerospace S&T Intelligence Studies Report Series Abstracts (5)*, 1995, pp. 288-303, *CAMA*, Vol. 3, No. 6; and Liu Jiaqi, "Penetration Technology for Tactical Missile Warheads," *Aerospace S&T Intelligence Studies Abstracts* (5), Vol. 95 (5), *CAMA*, 1996, Vol. 3, No. 6. Yang is from the Beijing Institute of Special Electromachinery (Beijing Teshu Jidian Yanjiusuo).

66. Huang Tung, "M-Series Missiles and New Navy Equipment—New Weapons in Taiwan Strait Exercises," *Kuang Chiao Ching*, April 16, 1996, pp. 22-25, in *FBIS-CHI*-96-097. The reporting on the EMP warhead is consistent with R&D underway on a HPM device, intended to target enemy electronic systems. For discussions on negating hardened targets, see Xu Xiaocheng, "Research on Penetration Depth of Projectiles Into Thick Concrete Targets," *Qiangdu Yu Huanjing*, 1996 (4), pp. 1-7, in *CAMA*, 1997, Vol. 4, No. 3. Xu is from CALT's 13th Research Institute. For a study addressing submunition dispersal problems, see Yan Dongsheng, "Technical Means for Reducing Dispersal of Mini-Warheads," paper presented at the October 1995 Annual Conference on Flight Mechanics, in *CAMA*, 1997, Vol. 4, No. 5.

67. General Second Artillery organizational information drawn from numerous sources, to include open source Chinese publications and from discussions while assigned as the assistant air attache in Beijing, China from 1992-1995. Also see PLA Directory of Personalities, USDLO Hong Kong, 1996, pp. 48-51; Bill Gertz, "New Chinese Missiles Target All of East Asia," *Washington Times*, July 10, 1997, p. 1; Hisashi Fujii, "Facts Concerning China's Nuclear Forces," *Gunji Kenkyu*,

November 1995, in *FBIS-CHI*-96-036; "2nd Artillery Honor Roll," *Flying Eagle (Changying)*, November 1993; "Outstanding Units," *Flying Eagle,* May 1992; Lewis and Xue, p. 213 footnote; and Nuclear Weapons Database, Vol. 5, pp. 324. Among sources, *Flying Eagle,* one of a handful of Second Artillery-associated publications, is most useful in piecing together the organizational structure.

68. *Ibid*. The tall white Second Artillery Headquarters building, which has the roman number "two" written on its surrounding walls, is clearly visible to the west when driving from Beijing through Qinghe on the way to the Great Wall at Badaling.

69. *Ibid*. Estimates vary wildly, ranging from 450 to 1200. China closely guards the size of its arsenal of warheads and missiles, and any estimate is only an educated guess. The 450 number is derived from Nuclear Weapons Database, Vol. 5, p. 324.

70. *Ibid*. This unit may also be known as the 51 Base. Base numbers (i.e., 51 Base) and types of missiles assigned are depicted in a 1996 Heritage Foundation graphic outlining regional missile threats. For details on DF-21, see Gertz, p. A16.

71. *Ibid*. Also see Richard D. Fisher, "China's Missiles Over the Taiwan Strait: A Political and Military Assessment," paper presented at September 1996 Coolfont Conference on the PLA, pp. 1-30. Huangshan is also known as Tunxi.

72. *Ibid*.

73. *Ibid*.

74. *Ibid*.

75. *Ibid*.

76. For information on Third Institute research into mapping, see Wang Yongming article in *Xiaoxing Weixing Jisuanji Xitong*, August 1995, in *FBIS-CST*-96-001. For a Third Institute targeting study, see Yuan Zaijiang, "Optimum Modeling Study and Algorithm of Targeting Selection," *Huoli yu Zhihui Kongzhi,* 1998, Vol. 23 (1), pp. 31-38, in *CAMA*, Vol. 5, No. 3.

77. This theory on the shift in Chinese nuclear doctrine is outlined in Alastair Iain Johnston, "China's New 'Old Thinking:' The Concept of Limited Deterrence," *International Security*, Winter 1995/1996, pp. 5-42.

78. Sun Zi'an, "Strategies to Minimize High Tech Edge of Enemy," *Xiandai Bingqi*, (*Modern Weaponry*), August 8, 1995, in *FBIS-CHI-96-036*; on the use of ballistic missiles against aircraft carriers, see Zhu Bao, "Discussion of Technical Means to Attack Aircraft Carriers With Tactical Ballistic Missiles," *Hubei Hangtian Keji*, 1997 (1), pp. 46-49, in *CAMA*, 1997, Vol. 4, No. 4. Zhu is from the 066 Base Institute of Precision Machinery.

79. "Selected Military Capabilities of the PRC," Department of Defense, April 1997, quoted in Timothy W. Maier, "U.S. Is Financing China's War Plan," *Insight*, May 12, 1997, p. 8.

80. Yu Guohua, "National Defense University Officer on Weaker Force Achieving Victory in Local War," *Zhongguo Junshi Kexue*, May 20, 1996, pp. 100-104, in *FBIS-CHI-96-252*. For studies on use of missiles against airfields, see Yu Renshun, "Research on Terminally Guided Submunitions For Blocking Airfield Runways," paper presented at November 1997 conference of National Missile Designers Network, in *CAMA*, Vol. 5, No. 3.

81. Lu Linzhi, "Preemptive Strikes Crucial in Limited High-Tech Wars," *Jiefangjun Bao*, February 14, 1996, p. 6, in *FBIS-CHI-96-025*.

82. Gu Baonan and Cai Xudong, "Rapid Maneuvering and Erecting Concept for Missile Launch Vehicles," *Space Launch Technology*, Vol. 92 (2), pp. 16-19.

83. Zhang Hu, "Application of GPS in Mobile Missile Positioning," Beijing Institute of Special Engineering Machinery paper presented at 1993 Conference on Launch Engineering and Ground Equipment, abstracted in *CAMA*, Vol. 1, No. 5.

CHAPTER 5

IN DEFENSE OF ITS OWN: CHINA'S AEROSPACE DEFENSE

China's quest for information dominance and growing competency in long-range precision strike is intimately related to its overall aerospace defense modernization effort. The PLA has long suffered from an antiquated air defense intelligence network, command and control, and capable air defense weapon systems. Specifically, the PLAAF has lacked decent early warning radars, early warning satellites, and automated intelligence handling and transmission facilities.[1] However, since the Gulf War, Chinese policymakers have placed a high priority on upgrading their overall air defense capability in order to protect strategically critical points. With China's space and missile industry leading the way, a comprehensive air defense network is being designed to counter air and missile attacks. The network is composed of numerous segments, including early warning, command and control, air defense weapons, and offensive counterair and counterspace weapons systems.

A national air defense reconnaissance network composed of a variety of sensors is a top priority. Chinese strategic analysts have consistently outlined the most important sensors for development. In addition to strategic HUMINT, missile early warning, photoreconnaissance, and SIGINT systems, priorities include airborne early warning platforms, bipolar/multipolar (*shuang/duodi*) radars, over-the-horizon sensors, ultrawideband (UWB), impulse (*chongji*), microwave, laser, millimeter, inverse synthetic aperture, and acoustic radars. A common theme throughout the literature is an outstanding requirement for an ability to counter stealthy aircraft and missiles.[2] China, especially the PLAAF, is clearly concerned about the effect of stealth weapons, such as the F-117A, on its air defense systems and is attempting to enhance its existing air defense system.[3] Crucial to its effort to form a integrated air defense network is the ability to fuse data from multiple sensors, introducing widespread data netting of sensors in order to reduce the value of stealth. Several institutes are engaged in working this problem.[4]

First, China is working to indigenously develop an airborne radar. In the meantime, however, the Chinese have contracted with Great Britain's Racal and Israel's IAI for the purchase of AEW radars. Racal will supply six to eight of its Searchwater surveillance radars to the PLA Navy. IAI, through Shaul Eisenberg's UDI office in Beijing, is negotiating the sale of the Phalcon radar which will be fitted into the Russian A-50 (IL-76 derivative). The contract is valued at 250 million U.S. dollars.[5] Aerostats are another type of airborne sensor being closely examined by the PLA Air Force and Navy. An aerostat (*qiqiu*) network, helpful for detection of low flying cruise missiles, would consist of a series of large inflated dirigibles hoisted above China's coastline.[6]

China's development of ultrawideband (UWB) radar technology is focused on counterstealth (*fan yinshen*). Proponents claim UWB (*chaokuandai*) systems can decrease radar cost, detect small, stealthy targets, defeat enemy jamming, and minimize the size of radars. UWB radars send repetitive transmissions of a single powerful pulse of a very wide bandwidth. Chinese experimental tests have been conducted against coated targets. UWB radars commonly have a bandwidth of up to one gigahertz and can track hundreds of targets.[7]

There are numerous other sensor programs underway, such as bistatic and multistatic radar development. A well-known counterstealth system, bistatic and multistatic radars use different sites for transmission and return of radar pulses.[8] Laser radars are another priority area of development. The Chinese are experimenting with radar-like devices called lidars, which use the laser light reflected from targets and received by an optical lens to locate the targets. A deliberately widened beam is first used to acquire a target and then reduced to a pencil beam in order to improve target calculations. Chinese researchers have emphasized CO_2 lasers in particular. The Chinese have also conducted research on a higher-power laser radar similar to the U.S. "Firepond," which has space tracking capabilities.[9] Engineers are also carrying out R&D into gigawatt-level high powered microwave as a counterstealth measure.[10]

China is also stepping up investment into developing an ability to counter antiradiation missiles through the use of low probability of intercept radars (*di zaihuo gailu leida*), phased array radars, and radar networking. The low probability of intercept (LPI) radar technology is concentrating on low peak power transmission and spread spectrum (*guangpin*) technologies. In addition, when radars of the same frequency form a net, they can cause deviation errors in the ARM guidance set. Signal adjustments can be made to ensure the signals operating on the same frequency do not interfere with each other. Widespread use of decoys and traps (*youpian*) to deceive ARM sensors is another option under close scrutiny.[11]

These sensors are critical for cueing kinetic and directed energy weaponry for air defense. Like many developing nations, China's air defense architecture is dominated by its ground based segment. PLAAF fixed wing assets remain the weak link in guarding China's airspace.[12] With the Second Academy in the lead, China is aggressively modernizing its surface-to-air missile capability by upgrading current systems; fielding new generation SAMs; and procuring foreign systems. Standard scenarios drawn up by Chinese analysts include initial launches of foreign tactical ballistic and cruise missiles against weak points in China's air defense structure, destroying air defense assets located along key air routes. Then an adversary, under an ECM shield, is expected to launch ARMs and precision air-to-ground munitions using combinations of aircraft and cruise missiles to attain air superiority.[13]

To counter an enemy air and missile campaign, PLA strategists and Second Academy analysts believe they must develop multicapable air defense systems able to engage a variety of missiles and aircraft at low to high altitudes. SAM systems must use advanced search, discrimination, tracking, and guidance technology; advanced mobile equipment, high quality materials, and powerful propulsion and fire control systems, together with multisensor warheads and advanced software. Chinese analysts stress the importance of an air defense "firepower network," comprised of SAMs, AAA, electromagnetic shells (*diancipao*), and air defense lasers.

In its SAM development, the Second Academy is striving for ever faster speeds and more effective guidance systems. Among other achievements, the Second Academy believes they have made significant advances in developing a passive long-wave infrared imaging terminal guidance seeker, which can be used in all weather conditions and is highly resistant to electronic countermeasures.[14] Engineers are pressing for completion of preliminary research into millimeter wave and infrared integrated seeker for hypervelocity missiles, similar to the U.S. ERINT system. Borrowing from foreign experiences, Chinese engineers are attempting to correct problems they have encountered.[15] In order to ensure the ability of future SAMs to operate in a the face of heavy jamming, CASC has designed digital simulators to allow operators to train in a dense electronic combat environment.[16]

China is taking advantage of every opportunity to incorporate foreign technology in order to hasten developmental objectives. Chinese clandestine acquisition of Patriot technology after the Gulf War, and the purchase of Russian S-300PMU (SA-10B) and S-300/PMU-1 (SA-10C) SAM systems have made important contributions to the Second Academy's ability to develop a highly capable indigenous SAM. One candidate for incorporating S-300 and PAC-2 Patriot technology is the HQ-9, which may have rudimentary missile defense capabilities, including defense against cruise missiles.[17] The Second Academy is probably most interested in phased array radar and missile guidance technology used in both the PAC-2/PAC-3 and S-300. China and Russian negotiators have also recently signed an agreement on the sale of TOR-M1 (SA-15) SAM systems.[18]

The Second Academy is carrying R&D into more advanced air defense technologies. For example, their systems engineering department has examined the utility of HPM weapons as an air defense weapon both to jam or destroy incoming aircraft.[19] Some call HPM weapons a "natural enemy" (*tiandi*) of more technologically advanced militaries and an "electronic trump card" (*dianzi shashou*), which can be used as counterstealth as well as an antiradiation weapon.[20]

Possibly related to HPM technology, more exotic air defense concepts under development include an electromagnetic missile (*dianci daodan*). This "missile" is a burst of radio frequency (RF) energy which can severely disrupt the electronic systems of an attacking aircraft or missile. Both the Second Academy and the University of Electronic Science and Technology of China have active electromagnetic missile programs which also have useful properties for counterstealth and long-range communications.[21]

Another relatively mature technology under development is the electromagnetic and electrothermal gun. China's first experimental railgun (*daoguipao*), developed by the China Academy of Engineering Physics, was tested in 1986, when a 0.34 gram object was hurled at 1.68 kilometers a second. China's first electrothermal gun (*dianrepao*) was tested in 1991 and achieved a velocity of 1.86 kilometers a second. CAS's Institute of Particle Physics also developed a railgun in 1988 which propelled a 50 gram shell at 3 kilometers a second. The same institute tested a coilgun (*dianquanpao*), capable of projecting a 44 kilogram object 14 meters per second, in 1990. CAEP developed its own electrothermal gun in 1991 and fired a 20 gram projectile at 1.85 kilometers a second.[22]

Finally, analysts advocate upgrading China's existing antiaircraft artillery (AAA) to meet challenges posed by enemy missiles, especially to counter the cruise missile threat. Modeling has been conducted against the BGM-109C Tomahawk, and the results gave 37mm AAA a kill rate of 23.0 percent. The air defense system was able to detect the cruise missile at a range of 43 kilometers and was able to track it at a range of 9.1 kilometers.[23]

Chinese strategists also stress limiting an adversary's information. Advocates urge adopting measures such as camouflage, deception, dispersal, mobility, and secrecy. Of great importance is physical destruction of an adversary's reconnaissance platforms,[24] employment of low probability of intercept radars and communications equipment, deployment of baits and false targets, and reduction the radar cross section (RCS) of air defense weapon systems. When not employed in combat, weapon systems must be hidden in bunkers or covered with camouflage netting, and emissions must be controlled and frequencies kept secret. Under combat conditions, air defense operators will theoretically plug into C^3I systems for targeting data and rapidly change positions, and be on guard against enemy special forces.[25]

Chinese operational concepts, however, indicate that Beijing believes it must never get to the point where enemy air assets are crossing into Chinese airspace. In a phased approach to air defense, first and most important is to destroy opposing force's air and missile assets before they leave the ground. According to aerospace analysts, the top priority in emerging air defense doctrine are preemptive strikes (*xianfa zhiren*) against adversarial air assets on the ground, especially high value assets such as airborne warning and control, and electronic reconnaissance platforms. Highly accurate ballistic and cruise missiles are crucial for preemptive strikes against airfields and supporting C^4I assets. Chinese writings have specifically noted the threat of UAVs and have strongly advocated striking out at nodes within the UAV C^3I system.[26]

After reducing the punch an adversary can deliver, Chinese analysts believe they must prepare key targets, such as C^3 centers, ports, airfields, missile sites, and UAV bases to counter enemy attacks. Attacking aircraft and cruise missiles must be identified as soon as possible. When outside of 100 kilometers, fighter aircraft and long-range SAMs must be employed. Between 10-100 kilometers, mid- to short-range SAMs, railguns, and air defense lasers must be activated. Once inside 10 kilometers, air defenders must rely on short-range and ultrashort-range SAMs and AAA.[27] Formation of composite air defense units is a key step in seamless air defense operations.[28]

Missile Defense. China's air defense concepts include protection against cruise and ballistic missiles as well. The

Second Academy and the Shanghai Academy of Spaceflight Technology are leading the way in China's TMD R&D. Western open source reporting and Chinese technical journals provide clear indication that the CMC has approved funding for a 10-year developmental program for a theater missile defense system, to include satellites for missile launch warning. The PLAAF and CASC have advocated a 15-year, three-phase approach to TMD development, first by fielding a "Patriot-like" system, such as the HQ-9; initiating model R&D on an extended range anti-theater ballistic missile modeled on the U.S. PAC-3 interceptor; and carrying out preliminary research on a THAAD-class system.[29]

Studies indicate the Second Academy has already initiated programs to develop means to counter tactical as well as medium-range ballistic missiles, focusing on infrared seeker development, and on combined infrared and millimeter-wave seekers. China is modeling and conducting computer simulation of a theater missile defense system that can counter missiles with a range of 2,500 kilometers. After successfully developing a missile defense system, initial concepts indicate they would cover major cities and operational centers of gravity, to include key ballistic missile units, and would also be organic assets of certain group armies.[30]

Chinese strategists and engineers recognize that a space-based early warning capability is essential to a viable missile defense architecture. China has a well-established technology base in infrared sensors, which, when placed on satellites, can detect a missile almost immediately after launch by detecting the infrared radiation from its engine or motor plume.[31] Technical writings indicate the space industry is working to master specific technologies associated with missile early warning satellites. In a potentially related program, CAST has primary responsibility for providing the satellite bus for the infrared telescope which, according to design outlines, will be placed in a geosynchronous orbit shortly after the turn of the century.[32]

Command and Control. China understands the problems of orchestrating an air defense campaign which has reconnaissance, offensive counterair, and defense counterair components. An integrated national air defense system is the long-term objective. The PLA Air Force has responsibility for air defense of Chinese territorial air space. Offices of the PLA Air Force leadership are located in the PLA Air Force headquarters complex in western Beijing. Similar to its U.S. counterparts, the headquarters oversees the organization, training, and equipping of the PLAAF. Headquarters, PLAAF also manages a command center which uses a semi-automated C^2 system to monitor airspace inside Chinese territorial airspace and maintain links with PLAAF units throughout the country. However, as a GSD service arm, PLAAF operational orders originate from the GSD Operations Department complex in the Western Hills.

The next echelon resides at the military region level. The various Military Region Air Force (MRAF) headquarters support PLA Military Region commanders by providing air defense and air support to ground operations. Each MRAF oversees one or more air corps or command posts which have operational command over aviation and air defense assets. Each command post/air corps is responsible for air defense in its particular area of operations.[33]

In a wartime situation, the Sector Operations Center (SOC) is C^2 authority over defense counterair assets within its area of responsibility. The SOC is being designed to receive data from over the horizon and missile early warning radars, and other sensors. It has authority to direct PLAAF and Navy air defense fighters, SAMs, AAA, and air defense jammers. The SOC also maintains control over any mobile tactical air defense systems which control radars, air defense fighters, SAMs, and AAA. The SOC reports to the regional Air Defense Operations Center.[34]

The key entity responsible for R&D and systems engineering design of China's automated C^2 systems is the Ministry of Electronics 28th Research Institute in Nanjing. The 28th Research Institute was a key player in the development of China's area air defense network, which acquires, processes, and represents the air situation for the area under its command. Naturally, the 28th Institute is also responsible for development of China's national air traffic control network.[35]

Counterspace. Another important aspect of China's aerospace defense program is found within the rubric of counterspace. Chinese writings have long given lip service to the need to gain the upper hand in space as a prerequisite to battlespace dominance. However, since the Gulf War, a growing body of literature from the Academy of Military Sciences (AMS), COSTIND, and China Aerospace Corporation (CASC) has advocated the development of a counterspace capability, and indicates ongoing programs to master enabling technologies.

Counterspace operations are viewed as an inevitable aspect of future warfare and as part of an overall information denial doctrine. AMS writings note that the United States relies on satellite platforms for 70 percent of its communications (90 percent for navy communications), and 90 percent of its intelligence.[36] Chinese strategists and engineers perceive U.S. reliance on communications, reconnaissance, and navigation satellites as a potential "Achilles' heel." COSTIND advocates believe China must develop space combat systems which are a fundamental aspect of the revolution in military affairs and a new sphere of warfighting.[37]

The PLA and the defense industries have implemented or are examining a broad range of passive and active counterspace measures. The PLA currently stresses passive counterspace measures in an attempt to deny foreign reconnaissance satellites with information on its disposition of forces and R&D programs. For example, in 1992 COSTIND and CASC established camouflage standards for missile development in order to counter foreign optical, infrared, and radar satellite systems.[38] Active measures under consideration include kinetic kill vehicles, ground-based lasers, and satellite jammers. CASC is among the strongest advocates for an antisatellite (ASAT) development program, arguing that an ASAT (*fanweixing*) capability is needed to deter potential adversaries from attacking China's own satellites. They also believe that an ASAT capability is easier to develop than ballistic missile defense systems. In fact, components of the two systems are intimately related.[39]

Preliminary research (*yuxian yanjiu*) on ASATs has been carried out since the 1980s, at least partly funded under the 863 Program for High Technology Development. However, R&D on fundamental technologies applicable to an ASAT weapons system have been ongoing since the 1960s. Under the 640 Program, the space and missile industry's Second Academy, traditionally responsible for SAM development, set out to field a viable antimissile system, consisting of a kinetic kill vehicle, high powered laser, space early warning, and target discrimination system components. While this program was abandoned in 1980, engineers associated with this effort are still active. For example, the 640 Program chief designer, Song Jian, is chairman of the State Science and Technology Commission.[40]

Chinese strategists recognize the importance of a space tracking network in counterspace operations. COSTIND is modernizing and expanding its space tracking network. This network, operated by COSTIND's China Launch and Tracking Control (CLTC), will be necessary for tracking and control of a projected increase in China's domestic satellites and its international satellite launch business. COSTIND is adding overseas links in Chile and the South Pacific island of Kiribati, and has contracted with France for access to data from its space tracking network.[41] China Academy of Sciences' astronomical observatories in Nanjing and Kunming feed into the CLTC network, providing orbital prediction data for CLTC. CAS and CLTC are upgrading their network of high resolution telescopes, augmented by laser tracking devices. China's space community claims an ability to detect objects in space down to 10 inches.[42] While the network is designed for cooperative targets, it does provide the framework for improvements against non-cooperative targets.

CASC researchers, primarily concentrated within the Second Academy, are leading China's preliminary and model R&D of counterspace systems. Several articles have appeared in the Second Academy's journal, *Systems Engineering and Electronics*, which indicate they are tackling some specific problems in negating adversarial satellites. Almost all deal with terminal guidance problems with a homing kill vehicle. Specific systems under evaluation and simulation include infrared, radar, and impulse radar terminal guidance.[43] Chinese engineers have also conducted studies to counter satellite decoys as well.[44]

Other entities associated with CASC have supported the overall R&D effort. Harbin Institute of Technology and Beijing University of Astronautics and Aeronautics, for example, have carried out modeling and simulation of various space intercept control and terminal guidance systems. One control concept involved the use of several small solid motors for orbital control stabilization.[45] CASC analysts within the Third Academy have conducted assessments of air launched ASAT missiles and have noted this approach as the most inexpensive ASAT option. Supporting assessments have focused on solid fueled motors for the air-launched miniature homing vehicle.[46]

Chinese aerospace analysts view ground-based high-powered lasers, able to degrade or destroy satellites at all altitudes, including medium and geosynchronous orbits, as an alternative to kinetic kill vehicles. Directed energy ASAT weapons are touted as the wave of the future.[47] China has the basic technologies needed to move to more advanced R&D stages of a ground-based laser (GBL) ASAT system. One of the most capable lasers under development is a free electron laser (FEL). COSTIND's China Academy of Engineering Physics (CAEP) began FEL (*ziyou dianzi jiguang*) development in 1985 and activated their first experimental system, the Shuguang-1, in May 1993. The experimental Shuguang-1, developed by CAEP's Southwest Institute of Fluid Physics, achieved a power output of 140 megawatts during a test in 1994. CAEP is working to reduce FEL size as a means to increase mobility. In addition to a FEL system, CAEP and CAS have made significant achievements in chemical and solid state lasers.[48] PLA-affiliated publications assert that while the Chinese do not yet possess the capability to destroy satellites with high-powered lasers, they are capable of damaging optical reconnaissance satellites.[49]

As an integral part of its high-powered laser program, China has placed high priority on development of adaptive optics and deformable mirrors. Adaptive optics, which use numerous electronic devices to shape a mirror for optimal beam pattern, are necessary to compensate for atmospheric absorption, scattering, and thermal blooming. During the 1994 COSTIND S&T Committee Meeting, "flexible mirrors" (*lingjing*) were singled out as a critical technology for increased emphasis. Concentrated effort into adaptive optics began in 1980 when the CAS formed an adaptive optics research group and laboratory within its Institute of Optoelectronics in Beijing. Since then, other CAS institutes, including Shanghai Institute of Optics and Fine Mechanics and Anhui Institute of Optics and Fine Mechanics, have lead China's adaptive optics R&D.[50] CAEP's Institute of Applied Physics and Computational Mathematics (IAPCM) has supported the effort through modeling work on atmospheric effects on ground based high-powered laser weapons.[51]

One other directed energy concept which the Second Academy has evaluated, and advocates, is the use of HPM weapons in a counterspace role.[52] AMS and COSTIND analysts believe HPM weapons will serve a useful purpose in the 21st century and strongly advocate their development. CAEP's Institute of Applied Electronics, University of Electronic Science and Technology of China, and the Northwest Institute of Nuclear Technology in Xian are three of the most important organizations engaged in the research, design, and testing of Chinese HPM devices.[53]

There have been other ASAT concepts alleged to be under consideration as part of an overall aerospace defense system. References, albeit dubious in nature, outline simple ASAT measures including an antisatellite satellite in a counter-orbit which releases 30 gram steel balls in order to penetrate the satellite shell. This option, called "fairy spreading flowers" (*xiannu san hua*), is low cost and effective. Another option specifically directed against space-based laser platforms calls for the application of powder, paint, or dust to cover reflectors in order to prevent laser weapons from focusing on targets. This measure is simple, but vulnerable to attack from enemy satellites.[54]

One final counterspace measure under examination is an electronic countermeasure directed against enemy satellites. COSTIND and CASC are conducting feasibility studies on various satellite jammers intended to complicate use of communication satellites and NAVSTAR GPS.[55] Such measures would deny an adversary use of a satellite, but not destroy the platform itself, perhaps avoiding escalation of the conflict.

There are some within China who are more circumspect about ASAT development and the militarization of space. Arms control experts, such as Du Xiangwan, have openly called for a international ban on ASATs and establishment of an international regime for oversight and control of military satellites. He is critical of certain missile defense systems, such as the space based laser, which has the capability to attack satellites as well as missiles.[56] Other Chinese arms control advocates see the development of ASATs as inevitable due to the growing importance of satellites for military operations. Therefore, limitations are required on military satellites as well as on ASAT development.[57]

The development and deployment of an active counterspace weapon system would require a decision at the highest level of the PLA leadership. The literature does not provide any clear indication that the CMC has directed the defense industrial complex to move toward ASAT testing or production stages. Technical writings do, however, clearly point to conceptual assessments on various ASAT systems and related technologies. AMS writings provide the doctrinal framework. The literature suggests China's ASAT program is beyond preliminary research and has moved into the model definition stage of R&D. These studies could form the basis of any joint CMC-State Council decision to proceed to the next step in the R&D cycle.

Cost would be a major factor in deciding whether or not to deploy an ASAT system. According to some estimates, basic space sensors, facilities, and manpower would run a fixed cost of approximately $20-30 million. Cost of on-board sensors ($1 million) and the missile ($1 million) would add up to a variable cost of around $2 million. If a facility launched 10 interceptors, its fixed and variable costs would be about equal, and the total cost per launch would be about $4-6 million per interceptor. Against a high performance satellite valued at $1 billion, such an interceptor would have a cost effectiveness ratio of about 200:1. In short, development of an ASAT capability could be determined to be an economically sound decision.[58]

There are a number of detractors, however, which may discourage the PLA from attempting to deploy a ground-based launcher (GBL) ASAT. First is the technical difficulty of atmospheric absorption, scattering and thermal blooming of the high-power laser. Further development of adaptive optics and choice of laser will help to mitigate this problem. Deuterium fluoride and free electron lasers can operate at wavelengths which minimize atmospheric problems. Basing a GBL on a high mountain, above cloud cover, can also reduce the amount of atmosphere through which a laser must travel. Basing a GBL in a relatively arid region, such as Gansu, Qinghai, or Xinjiang would help reduce weather problems.

Another detractor is the infrequency in which a low earth orbit reconnaissance satellite would pass within view of the GBL site. Deployment of a larger number of GBLs would increase opportunities, but would also drive up the costs. One other factor is the vulnerability of a GBL. A static GBL tucked away inside a mountain could be concealed, especially if potential adversaries were not looking for it. However, after its use on a satellite, the GBL would become a primary target for adversarial strikes.

Absent from discussions of active counterspace systems are clues as to the type of launcher which could be used to deliver a homing vehicle to its designated target. Chinese writings indicate that the former Soviet Union utilized the SS-9, which could counter recon, navigation, and weather satellites, as well as the space shuttle. They also detail an unidentified U.S. three-stage ground launched solid direct ascent ASAT (*zhijie shenggao fanweixing*) and the air-launched ASAT missile. Analysts, however, are critical of shortcomings of the Russian system in the face of U.S. satellite maneuverability, radiation and laser hardening, and advanced warning systems; as well as the 2,000 kilometer altitude limitations of the Russian coorbital ASAT (*gongguishi fanweixing*) system, and amount of time (at least 3 hours) it would take to reach its target. CASC analysts are also critical of the Russian system, citing the requirement to be tied to satellite launch centers. On the other hand, a fighter armed with an ASAT missile is praised for its flexibility and mobility.[59]

China's defense industrial complex clearly has set its sights on an integrated national air defense system, together with a counterspace capability, that would be able to oppose any would be attacker. The Second Academy's role in planning, doctrinal development, and systems R&D is significant should not be underestimated. The adversary used for its threat based planning, aquisition, and development is clearly the United States. The Second Academy, COSTIND, and other PLA entities carefully studied how the United States decimated the Iraqi air defense system and are taking steps to ensure China is capable enough to not suffer a similar fate.

ENDNOTES - CHAPTER 5

1. PLAAF deficiencies are discussed at length in Kenneth W. Allen, Glenn Krumel, and Jonathon Pollack, *China's Air Force Enters the 21st Century*, RAND Project Air Force study, 1995, pp. 112-113.

2. The Chinese believe this combination of sensors will be especially useful against cruise missile attacks. Key research institutes involved in radar development include the electronics industry's 14th Research Institute in Nanjing and CASC Second Academy's 23rd Research Institute in Beijing. See Xu Xingci, "Xunhang Daodan Fangyu Cuoshi" ("Cruise Missile Defense Measures"), *Zhongguo Hangtian*, April 1996, pp. 36-39. For a good comprehensive look at future directions in Chinese air defense, see Xu Xingju, "Tigao Woguo Dimian Fangkong Wuqi Dianzizhan Nengli" ("Raising China's Ground based Air Defense Electronic Warfare Capability"), *Xiandai Bingqi* (*Modern Weaponry*), August 1995. Xu, who has written widely on air and missile defense topics, is from the CASC's Second Academy Information Group. Also see "Counterstealth Ground Radars," *Hangtian* (*Spaceflight*), March 1997,

pp. 20-21, in *FBIS-CHI*-97-225, for an account of two radars. The first is an inverse SAR system which can image aircraft, warships, and satellites; and the second, the J-231, developed by the Second Academy's 23rd Research Institute, can detect and track targets at 30,000 meters. Both supposedly have an effective counterstealth capability.

3. Wang Xuichun, "Threat of Stealth Weapons to Modern Air Defense," *Xiandai Zhanshu Fangkong Jishu Wenzhai (Modern Tactical Air Defense Technology Abstracts)*, October 1994, pp. 42-47, in *CAMA*, Vol. 3, No. 6; Hu Jian, "Research on Counterstealth Technology for the HQ-2 Surface-to-Air Missile System," *Air Force Missile Academy Journal*, 1994, Vol. 16, No. 3, in *CAMA*, Vol. 2, No. 5. Prominent U.S. advocates of stealth technology are skeptical that China, or any other country, would ever be able to counter the U.S. lead in stealth technology.

4. Wu Ji, "Multisensor Data Fusion Technology and Its Application," *Kongjian Dianzi Jishu*, February 1994, pp. 1-6, in *CAMA*, 1994, Vol. 1, No. 5; Yu Wenxian, "Comments on Multisensor Information Fusion," *Guofang Keji Daxue Xuebao*, 1994, Vol. 16, No. 3, in *CAMA*, Vol. 2, No. 1. Yu is from NUDT. Zhang Saijin, "Multisensor Data Fusion for Target Detection, Recognition, and Tracking," *Hongwai Yu Jiguang Gongcheng*, 1996, Vol. 25, No. 2, pp. 21-26, in *CAMA*, Vol. 3, No. 4.

5. Amnon Barzilay, "Russia About To Decide on Ilyushin for IAI-PRC Deal," *Ha'aretz*, September 24, 1996, p. A1, in *FBIS-NES*-96-186.

6. PLAAF S&T Department representatives were discussing aerostat options with foreign producers in 1993. Current status is unknown. One could surmise that since signing an AEW contract with Israeli and Russian firms, the interest in an aerostat system could have diminished.

7. He Jianguo, Lu Zhongliang, and Liu Kecheng, "Counterstealth Mechanism of UWB Radar," *Guofang Keji Daxue Xuebao (Journal of National University of Defense Technology)*, February 1997, Vol. 19, No. 1, pp. 71-76, in *FBIS-CST*-97-014.

8. Among numerous references, see, for example, He Weixing, "Position Method and Precision Analysis for Bistatic Radars and Its Network System," *Fourth Annual Exchange of the China Aviation Society*, May 1992, pp. 42-47, in *CAMA*, Vol. 2, No. 2; Liu Qi and Sun Zhongkang, "Analysis of Detection Range in Air-Ground Bistatic

Radar," *Guofang Keji Daxue Xuebao,* October 1997, pp. 14-17, in *FBIS-CHI*-98-007; Wu Xielun, "Development of Multistatic Radar Networking Technology and Its Application Against Stealth Vehicles," *Xiandai Fangyu Jishu,* January 1995, pp. 8-20, in *CAMA,* Vol. 2, No. 2; and Wu Peilun, "Development and Application of Bistatic/Multistatic Radar Network," *Astronautics Research Information Report of the Second Academy,* Vol. 2, November 1995, in *CAMA,* Vol. 4, No. 6.

9. Sun Baoju and Chen Gang, "CO2 Jiguang Leida Jiqi Yingyong"("Applications of CO2 Laser Radar"), *Zhongguo Hangtian,* February 1993, pp. 44-46; Mao Guo'an, "Angular Tracking System of Coherent CO^2 Laser Radar," *Xitong Gongcheng Yu Dianzi Jishu,* 1996, Vol. 18, Vol. 3, pp. 47-54, in *CAMA,* Vol. 3, No. 4; and Zhu Dayong, "Super-Long Range Doppler Laser Radar," *Hongwai Yu Jiguang Gongcheng,* 1996, Vol. 25, No. 1, pp. 8-15, in *CAMA,* Vol. 3, No. 4.

10. Liang Baichuan, "Research on Active Stealth Techniques," *Shanghai Hangtian,* September 29, 1997, pp. 12-16, in *FBIS-CHI-*97-272.

11. Zhang Running and Chen Guangfei, "Counter-ARM Signal and Its Doppler Phase Compensation," *Dianzi Kexue Xuekan (Journal of Electronics),* January 1, 1997, in *FBIS-CST*-97-012; Shi Zhen, "Digital Simulation Study on Deception Technology For Countering Antiradiation Missiles," *Danjian Yu Zhidao Xuebao (Journal of Projectiles, Rockets and Guidance),* February 1996, pp. 22-30, in *CAMA,* Vol. 3, No. 6. Shi is from Harbin Institute of Technology; and Wang Weilun, "Research on Spread Spectrum and Bistatic Radar Systems," *Zhongguo Hangkong Xuehui Leida Yu Zhidao Jishu Wenzhai (China Aviation Society Radar and Guidance Technology Abstracts),* May 1993, 7 pp., in *CAMA,* Vol. 2, No. 1; and Chen Zongfei, "Alarm of Antiradiation Missile Directed Against Tracking Radars," *Shanghai Hangtian,* 1995, Vol. 12, No. 3, in *CAMA,* 1995, Vol. 2, No. 5. Chen is from the Shanghai Institute of Electromechanical Engineering. Also see Shi Zhen, "Simulation Research on Defense Against Anti-Radiation Missiles By Phased Array Radars," *Daojian Yu Zhidao Xuebao,* 1997 (1), pp. 19-22, in *CAMA,* 1997, Vol. 4, No. 3.

12. For an overview of these problems, see Kenneth Allen, Glenn Krumel, and Jonathon Pollack, *China's Air Force Enters the 21st Century,* RAND: Santa Monica, 1995.

13. Xu Xingju; and Zang Jize, "Developmental Direction of China's Ground-Based Air Defense Systems" ("Wo Guo Dimian Fangkong Wuqi Xitong de Fazhan Fangxiang"), *Xiandai Wuqi,* 1995.

14. Passive imaging infrared sensors use the infrared (heat) energy emitted from targets as their detection mechanism. They differ from simple infrared sensors in that they build up a pixelated image of the scene and match it with a digitized template of the target. See Guo Jinzhou, "Signal Processing System for Infrared Imaging Seekers," *Hongwai Yu Jiguang Gongcheng*," August 1996, pp. 12-19, in *FBIS-CST*-97-005. The Second Academy's 25th Research Institute is leading the effort with assistance from COSTIND's National University of Defense Technology and MEI's 11th and 13th Research Institutes. Also see Li Qunzheng, "Xunsu Fazhan de Hongwai Chengxiang Jingque Zhidao Jishu" (Rapid Development of Infrared Imaging Precision Guidance Technology), *Zhongguo Hangtian*, February 1991. Engineers are most confident about their mercury cadmium telluride (HgCdTe) long wave infrared (LWIR) detectors, developed by the North China Research Institute of Electro-Optics (NCRIEO).

15. He Liping, "Research on Millimeter Wave/Infrared Composite Seeker in Hypervelocity Missile," unpublished paper by the Aerospace Institute on S&T Information, *CAMA*, Vol. 2, No. 3. Published in *Hongwai yu Jiguang Gongcheng*, 1996, Vol. 25 (4), pp. 56-65, in *CAMA*, Vol 3, No. 6; Liu Yongchang, "Analysis of IR/MMW Combined Seeker," *Hongwai Jishu*, 1994, Vol. 16, No. 4, pp. 1-8, in *CAMA*, Vol. 2, No. 4.

16. Zhou Shuigeng, "Design of Digital Simulator of Jamming Environment of a Surface-to-Air Missile System," *Shanghai Hangtian*, 1995, Vol. 12, No. 1, pp. 10-14, in *CAMA*, Vol. 2, No. 2.

17. David Fulghum, "China Exploiting U.S. Patriot Secrets," *Aviation Week & Space Technology Review*, January 18, 1993, p. 20.

18. "Moscow Expects to Increase Military Cooperation With China," *Moscow Interfax*, June 27, 1997.

19. Zhang Hongqi, "High Powered Microwave Weapons," *Xiandai Fangyu Jishu*, April 1995, pp. 38-46, in *CAMA*, 1995, Vol. 2, No. 5. Zhang is from Second Academy's Beijing Institute of Electronic Systems Engineering. HPM technology is discussed in greater detail in other areas of this report.

20. Gong Jinheng, "High Powered Microwave Weapons: A New Concept in Electronic Warfare," *Dianzi Duikang Jishu*, February 1995, pp. 1-9, in *CAMA*, 1995, Vol. 2, No. 5. Gong is from the Southwest Institute of Electronic Equipment (SWIEE), China's premier electronic warfare research entity.

21. Zhu Zhichi, "Future Wars and the Electromagnetic Missile," Astronautics Publishing House paper HQ-94029, 1994. Zhu is from the Second Academy's 207th Research Institute; also see "Preliminary Experimental Study of Electromagnetic Missiles Conducted," *Dianzi Keji Daxue Xuebao*, February 1992, in *JPRS-CST*-92-013; and "Study of Backscattering of Electromagnetic Missiles," *Dianzi Xuebao*, June 1992, in *JPRS-CST*-92-023. As a side note, one of China's foremost experts on HPM devices is Lin Weigan, who earned his PhD from UC Berkley in 1950.

22. Wang Ying and Xiao Feng, *Dianpao Yuanli (Electronic Gun Principles)*, Beijing: National Defense University Press, 1993, p. 9-11.

23. Zhu Jin and Tan Jianjun, "Simulation Research on 37-mm Twin Antiaircraft Gun System to Intercept Cruise Missiles," *Huoli yu Kongzhi*, September 1996, pp. 24-31, in *FBIS-CST*-97-002; and Zang Jize, "Wo Guo Dimian Fangkong Wuqi Xitong de Fazhan Fangxiang" ("Developmental Direction of China's Ground-Based Air Defense Systems"), *Xiandai Wuqi*, 1995.

24. There are clear indications CASC, with PLAAF support, is designing an anti-AEW air-to-air missile. Key institutes include the Shanghai Institute of Electromechancial Engineering and the PLAAF's 8th Research Institute. Xing Xiaolan, "Development of Anti-AWACs Weapon System," *Yuancheng Kongkong Daodan*, June 1996, pp. 14-21, in *CAMA*, Vol. 4, No. 2; Li Guangpu, "Long Range Multi-use Air-to-Air Missile: Primary System for Attacking Radiating High Value Targets,"*Yuancheng Kongkong Daodan*, June 1996, pp. 1-13, in *CAMA*, Vol. 4, No. 2; and Zhang Wangsheng, "Development of Long Range Multipurpose Air-to-air Missile and Construction of An Air Defense System," paper presented at the November 1996 China Astronautics Committee, Annual Meeting of Unmanned Aerial Vehicle Specialists, *CAMA*, Vol. 4, No. 4.

25. Xu Xingju, p. 27.

26. Liu Xuejun and Zhang Changliu, "Study of Measures to Counter Unmanned Aerial Vehicles," *Guoji Hangkong*, March 1, 1996, pp. 21-23, in *FBIS-CHI*-96-176; and Xu Xingju, p. 17.

27. See Xu Xingci, "Xunhang Daodan Fangyu Cuoshi," pp. 36-39; and Xu Xingju, "Tigao Woguo Dimian Fangkong Wuqi Dianzizhan Nengli."

28. Wan Mingjie, "Study of Composite Deployment of Surface-to-Air Missile Troops in Air Defense of Strategic Area," *Air Force Missile Academy Journal*, 1994, Vol. 16, No. 3, in *CAMA*, Vol. 2, No. 5.

29. Such an approach is consistent with the space and missile industry's "Three Moves on A Chessboard" development strategy to have three systems in the R&D and production cycle at the same time. Gao Fuli, "Development Strategy and Serial Research of Anti-Tactical Ballistic Missiles," *Foreign Missile Technology Development in 2000*, October 1994, pp. 48-59, in *CAMA*, Vol. 2, No. 4. The three phase approach (*sanbuzou*) for China's TMD development is also discussed in Yang Chunfu and Liu Xiao'en, "Research Study on U.S. Ballistic Missile Development Plan," Aerospace Information Paper HQ-96009, 1996, in *CAMA*, Vol. 4, No. 2. A PAC-3-class interceptor would likely require a millimeter wave, track-via-missile capability.

30. "Source on Developing Multistage Air Defense System," *Sankei Shimbun*, May 30, 1996, p. 1, in *FBIS-CHI*-96-107; and Zhu Zhenfu and Huang Peikang, "TBM IR Radiant Signature, Selection of Optimum Operating Band for Anti-Missile Defenses," *Xitong Gongcheng Yu Dianzi Jishu* (*Systems Engineering and Electronics*), January 1996, pp. 9-17. Zhu and Huang are from the Second Academy's 207th Research Institute. Also see Wen Xianqiao, "Parameter Modeling of Attack Countermeasures for Anti-Tactical Ballistic Missiles," *Xiandai Fangyu Jishu* (Modern Defense Technology), April 1996, pp. 11-23, in *CAMA*, Vol. 3, No. 6; and Chen Jianxiang, "Research on Counter-Parallel Interception of Tactical Ballistic Missiles," *Xitong Gongcheng Yu Dianzi Jishu*, 1998, Vol. 20(4), pp. 18-22, in *CAMA* Vol. 5, No. 5. Wen Xianqiao and Chen Jianxiang are from the Second Academy's Beijing Institute of Electronic Systems Engineering. Also see Li Guangpu, "Army-Level Antitactical Ballistic Missile System," *Shanghai Hangtian*, February 1994, pp. 36-38, in *CAMA*, Vol. 2, No. 1. Li is from Shanghai Institute of Electromechanical Engineering. For a discussion on endoatmospheric TMD development, see Wu Guanghua, Liu Yongchang, and Gu Yuquan, "Dual-Mode Millimeter-Wave/IR Seeker for Endoatmospheric Interception," *Danjian Jishu* (*Projectile and Rocket Technology*), 1996, Vol. 9, No. 3, in *FBIS-CHI*-97-261. Wu, Liu, and Gu are from the Xian Qinghua Electromechanical Technology Institute.

31. AMS and defense industry officials consistently advocate missile early warning satellites in concepts for a national reconnaissance network. Leading institutes for infrared detector R&D include Beijing Institute of Remote Sensing Equipment, Shanghai Institute of Technical Physics, North China Research Institute of Optoelectronics, Kunming Institute of Physics, and Shanghai Xinyue Instruments Factory. CAST's Lanzhou Institute of Physics is a key provider of

cyrogenic equipment for cooling the infrared sensors. See Liu Jintian, "Hongwai Qijian Guoneiwai Fazhan Dongtai" (Developmental Prospects of Chinese and Foreign Infrared Devices), *Zhongguo Hangtian*, March 1992, pp. 41-45; and Wu Runchou, "Hangtian Linghuo Hongwai Jishu de Fazhan" (Development of Space Infrared Technology), *Zhongguo Hangtian*, March 1993, pp. 19-23. For other references to China's space-based infrared/ultraviolet telescope designs, see Chen Longzhi, "New Developments in Space Cryogenic Optics," *Diwen Gongcheng* (Cyrogenic Engineering), March 1994, pp. 9-13, in *CAMA*, Vol. 1, No. 5; and Ma Pinzhong, "Woguo Kongjian Wangyuanjing Fazhan" (Development of China's Space Telescope), *Zhongguo Hangtian*, July 1994, pp. 29-32. To give China the benefit of the doubt, the above references do not explicitly state these infrared sensors will be used for missile early warning. Space-based infrared sensors could plausibly be used for scientific or meteorological purposes. However, China, like many other countries, downplays military uses of satellites, and would be expected to mask any early warning satellite behind the facade of a scientific platform.

32. The former Soviet space-based early warning system, completed in 1982, used a constellation of satellites in highly elliptical Molniya orbits. The first U.S. missile early warning system, MIDAS, was launched in the early 1960s and was followed by Defense Support Program (DSP) satellites in the early 1970s. DSP satellites use infrared telescopes, backed by an optical component, in a geosynchronous orbit. For information on Chinese missile early warning systems and associated technology, see Lu Mingyu, Yi Kui, Yang Junfa, and Deng Ruzhen, "Development of Signal Source for Real-Time Infrared Earth Sensor," *Zhongguo Kongjian Kexue Jishu*, June 1996, pp. 63-70, in *FBIS-CST*-96-016; and Qiu Yulun, "Staring Focal Plane Array Imaging for Missile Early Warning," *Kongjian Jishu Qingbao Yanjiu*, May 1995, pp. 150-160, in *CAMA*, 1997, Vol. 4, No. 2.

33. Kenneth W. Allen, Glenn Krumel, and Jonathon Pollack, *China's Air Force Enters the 21st Century*, RAND Project Air Force study, 1995.

34. This concept, part of an advertised Theater Air Defense System (TADS), is outlined in an undated brochure from the 28th Research Institute, Nanjing, obtained during author's January 1996 visit.

35. *Ibid.*

36. Zhu Youwen and Feng Yi, *Gaojishu Tiaojianxia de Xinxizhan* (*Information Warfare Under High Technology Conditions*), Beijing: Academy of Military Science Press, 1994, p. 316. The U.S. Navy *is*

inextricably dependent on satellite communications for its every mission from peacetime forward presence to war at sea.

37. Liang Zhenxing, "New Military Revolution and Information Warfare," p. 8.

38. Li Chunshan, "Introduction and Explanation of the National Military Standard 'Camouflage Requirements for Surface-to-Surface Missile Weapon Systems'," *Hangtian Biaozhunhua* (Space Standardization), 1994, Vol. 5, pp. 12-15, summarized in *CAMA*, 1995, Vol. 2, No. 1. Li is from the Beijing Space Systems Engineering Design Department.

39. Yang Chunfu and Liu Xiao'en, "Fanweixing Wuqi Xitong Fazhan Qianjing de Yanjiu" ("Study on the Developmental Prospects for ASAT Weapons"), *Aerospace Information Research*, HQ-93005, CASC 707 Institute.

40. See Lewis and Xue for information on the 640 program. As a side note, leading U.S. experts have noted that ABM systems generally have inherent capabilities as ASATs, but the converse is not always true.

41. Wang Chunyuan, *China's Space Industry and Its Strategy of International Cooperation*, Stanford: Stanford University, Center for International Security and Arms Control, July 1996, p. 4; "China Building Satellite Tracking Station on Tarawa," *Asian Defense Journal*, March 1997, p. 66; and "Satellite Command Station Operational in Kiribati," *Zhongguo Xinwenshe*, October 14, 1997, in *FBIS-CHI-97-287*.

42. Trip report, NASA visit to China, June 12-22, 1991. For example, China plans to develop a 500-meter aperture radio space telescope for deep space exploration. With a price of approximately 25 million dollars, the system, which will be based in Guizhou Province, will primarily support civilian academic research, but could also be used to supplement China's space surveillance network. CAST and the China Academy of Sciences are involved. See "Beijing Plans to Develop 500 Meter Radio Telescope," *Xinhua*, April 9, 1998, in *FBIS-CHI-98-099*.

43. Examples include Yin Xingliang and Chen Dingchang, "Guidance and Control in Terminal Homing Phase of a Space Interceptor," *Xitong Gongcheng yu Dianzi Jishu (Systems Engineering and Electronics)*, 1995, Vol. 17, No. 6, pp. 1-9, summarized in *CAMA*, 1995, Vol. 2, No. 5; Yin Xingliang, Chen Dingchang, and Kong Wei, "TESOC Method Based on Estimated Value Theory For a Space Interceptor in Terminal Homing," *Xitong Gongcheng yu Dianzi Jishu*, 1995, Vol. 17, No. 8, pp. 1-14; Yin Xingliang, Chen Dingxiang, and Yuan

Qi, "Systems Design for Terminal Homing and Option for Lateral Acceleration of Space Interceptor," *Xitong Gongcheng yu Dianzi Jishu*, 1995, Vol. 17 (5), pp. 1-10, in *CAMA*, Vol. 2, No. 4; and Shi Xiaoping, Wang Zicai, and Ke Qihong, *Acta Aeronautica et Astronautica Sinica*, 1995, Vol. 16, No. 3, pp. 291-298. Also see Chen Haixin, "Longwave Infrared Automated Target Recognition System For Space Target Seekers," *Hongwai yu Jiguang Gongcheng* (Infrared and Laser Engineering), 1996, Vol. 25 (3); Li Fengchun, "Summation of the Micro-Orbital Control Rocket Motor For the Miniature Homing Interceptor," *Guti Huojian Jishu (Solid Motor Technology)*, 1994, Vol. 18, No. 2, pp.1-7, in *CAMA*, Vol. 1, No.5; and Zhou Hongjian, Zhao Yongjun, and Wu Ruilin, "Application of PWPF Modulator in Space Interceptor Side Force Control," *Xitong Gongcheng yu Dianzi Jishu*, April 1997, pp. 9-12, in *FBIS-CHI*-97-272. For reference to infrared terminal guidance, see Shi Xiaoping, "Estimation of the Range Between the Intercepter and Target During Infrared Terminal Guidance of Space Interception," *Hangkong Xuebao*, 1995, Vol. 16, No. 3, pp. 281-298, in *CAMA*, 1996, Vol. 3, No. 5. Chen Dingchang, obviously very involved in evolving ASAT technology, is director of the CASC Second Academy and editor-in-chief of *Systems Engineering and Electronics*. Impulse radar is also known as ultrawideband radar, which Chinese engineers see as one of several approaches to counter stealth technology. Some of these references could be refering to exoatmospheric intercept of ballistic missiles.

44. Xu Hui and Sun Zhongkang, "Temperature Differences Between Satellites and Satellite Decoys," *NUDT Journal*, 1994, Vol. 16, No. 3; also see Li Hong, "Identification of Satellites and Its Decoys Using Multisensor Data Fusion," *Xiandai Fangyu Jishu*, June 1997, pp. 31-36, in *CAMA*, Vol. 5, No. 1. Li is from the NUDT Electronic Technology Department.

45. For references to control problems, see Deng Zichen, "Problems in High Precision Computation for Nonlinear Control of Space Interceptors," *Feixing Lixue*, 1998, Vol. 16 (1), pp. 85-89, in *CAMA*, Vol. 5, No. 5. Yang Yingbo, "Control Research on a Space Interceptor in the Terminal Guidance Phase," unpublished BUAA paper, May 1994; Shi Xiaoing, "Study on Pulse Guidance Law for Space Interception," in *Zhidao yu Yinxin*, 1994, (4), pp. 1-4, in *CAMA*, Vol. 2, No. 3. Shi is from the Harbin Institute of Technology's Simulation Center. Deng is from Northwest Polytechnical University. For other studies, see Li Zhongying, "Study on Mid-Course Guidance for Aerodynamic Control of Anti-Missile Defense," unpublished paper (BH-B4774), Beijing University of Aeronautics and Astronautics (BUAA), May 1996, in *CAMA*, Vol. 5, No. 5; and Li Zhongying, "Approximative Estimation of Optimal Guidance for Frontal Ballistic Missile Intercepts," unpublished

BUAA paper (BH-B4776), in *CAMA*, Vol. 5, No. 5; and Li Zhongying, "Mathematical Modeling of Optimal Guidance for Anti-Tactical Ballistic Missiles," unpublished BUAA paper (BH-B4854), May 1996, in *CAMA*, Vol. 5, No. 5.

46. Yang Qunfu and Liu Xiao'en, "Fanweixing Wuqi Xitong Fazhan Qianjing de Yanjiu" ("Study on the Developmental Prospects for ASAT Weapon Systems"), *Hangtian Qingbao Yanjiu (Aerospace Information Research)*, March 1993. According to one footnote, the Third Academy's Cheng Delin was responsible for an assessment of an air launched ASAT system. The footnote was associated with a portion of the paper which addressed China's domestic ASAT program. Unfortunately, censors omitted this part of the study. Also see Li Fengchun, "Summation of the the Miniature Homing Interceptor and Its Micro Orbital Control Solid Motors," in *Journal of Solid Rocket Motor Technology*, 1994, Vol. 18 (2), in *CAMA*, Vol. 1, No. 5.

47. Yang and Liu, p. 77. See Appendix IV for more on China's directed energy weapons program.

48. Hui Zongxi, "Research, Design of SG-1 Free Electron Laser," *Qiang Jiguang yu Lizishu*, August 1990, in *JPRS-CST*-91-014; "COSTIND Minister on Aerospace, Laser Advances," *Renmin Ribao*, April 3, 1996, in *FBIS-CHI*-082-96; and "SG-1 FEL Amplifier Output Reaches 140MW," *Qiang Jiguang yu Lizishu*, December 1994, in *JPRS-CST*-94-019. On FEL miniaturization, see "High Performance Short Period FEL Wiggler," *Qiang Jiguang yu Lizishu*, May 1992, in *JPRS-CST*-92-019. The output power probably does not represent the actual amount of energy which would be delivered to the target.

49. Chou Kuan-wu, "China's Reconnaissance Satellites," *Kuang Chiao Ching*, March 16, 1998, pp. 36-40, in *FBIS-CHI*-98-098. *Kuang Chiao Ching*, or *Wide Angle*, is a Hong Kong-based publication with close ties to the PRC military establishment. Official U.S. Government reports are consistent with this assessment. The 1998 Report to Congress on PRC Military Capabilities (pursuant to Section 1226 of the FY98 National Defense Authorization Act) states "China already may possess the capability to damage, under specific conditions, optical sensors on satellites that are very vulnerable to damage by lasers. However, given China's current interest in laser technology, it is reasonable to assume that Beijing would develop a weapon that could destroy satellites in the future."

50. See "Qian Xuesen at COSTIND S&T Committee Meeting," *Keji Ribao*, March 14, 1994, p. 1; "Nation's Adaptive Optics Technology in World's Front Ranks," *Zhongguo Kexue Bao*, May 21, 1991, in

JPRS-CST-91-015; "Adaptive Optics Technology Is World Class," *Keji Ribao*, October 30, 1992, in *JPRS-CST*-92-025; "Effects of Deformable Mirror/COAT System Finite Subaperature Size on Compensation Efficiency," *Zhongguo Jiguang (Chinese Journal of Lasers)*, February 1992, in *JPRS-CST*-92-010; and "Demonstration of Uniform Illumination on Target Focusing High Power Laser Beam," in *Guangxue Xuebao* (Acta Optica Sinica), March 1992, in *JPRS-CST*-92-013.

51. "Methods for Investigating Near-Field Power in Ground-Based High Power Laser Weapon Testing," *Zhongguo Jiguang (Chinese Journal of Lasers)*, December 1992, in *JPRS-CST*-93-008.

52. Li Hui and Wang Zibin, "Development of High Powered Microwave Weapons and Their Applications for Counterspace and Air Defense," *Zhidao yu Yinxin*, 1995 (1), pp. 3-15, in *CAMA*, 1995, Vol. 2, No. 3. This article has also appeared as an *Aerospace Information Research* article, HQ-94030. China's HPM program is discussed in more detail in another part of this study.

53. Zhu Youwen and Feng Yi, *Gaojishu Tiaojianxia de Xinxizhan*, (Information Warfare Under High Technology Conditions), Academy of Military Science Press, 1994, pp. 308-310; "Beam Energy Weaponry: Powerful as Thunder and Lightening," *Jiefangjun Bao*, December 25, 1995, in *FBIS-CHI*-96-039; "Outlook for 21st Century Information Warfare," *Guoji Hangkong*, (*International Aviation*), March 5, 1995, in *FBIS-CHI*-95-114; "Microwave Pulse Generation," *Qiang Jiguang yu Lizishu*, May 1994, in *JPRS-CST*-94-014. Most writings on HPM development come from these three organizations.

54. Wang Yongkui, "Can China Successfully Protect The Three Gorges Dam?," *Tangtai*, October 15, 1993, in *FBIS-CHI*-93-208.

55. Wu Jinliang, "Range Testing of Satellite Communication Countermeasures," *Electronic Countermeasure Technology and Intelligence Reform Abstracts*, November 1995, pp. 96-101, in *CAMA*, 1996, Vol. 3, No. 6. Reference to a Chinese study on a GPS jammer is included in author's unpublished report, *China's Space and Missile Industry*, June 1995.

56. Du Xiangwan, "Analysis and Discussion of Arms Control in Space," paper presented at the bilateral meeting between the Chinese People's Association for Peace and Disarmament and the U.S. National Academy of Sciences, October 16-18, 1992 in Beijing. Du Xiangwan, Deputy Director of the Institute of Applied Physics and Computational Mathematics, is China's leading expert on arms control issues.

57. Liu Erxun, "Arms Control in Outer Space," paper presented at the same bilateral conference cited in the previous footnote. Liu is an arms control expert from CASC's China Academy of Launch Technology.

58. This assessment is outlined in detail in *New World Vistas: Air and Space Power for the 21st Century*, (Space Applications Volume), p. 56.

59. Yang and Liu, p. 76.

CHAPTER 6

CONCLUSION

China's military industrial complex—driven by an evolving doctrine which emphasizes information dominance, preemptive, long-range precision strikes against critical nodes, and a highly capable air defense—is conducting preliminary and applied R&D into a dizzying array of technologies and weapons systems. China's success, however, in achieving technological breakthroughs and fielding the kind of reconnaissance/strike complex discussed in this monograph is not assured. Yet even modest breakthroughs could advance the PLA's ability to achieve its national security objectives, including forceful integration of Taiwan into the PRC, and deny the United States an ability to intervene.

Obstacles in PLA Strategic Modernization.

There are a number of obstacles which will complicate China's ability to modernize the PLA. First, defense R&D and production must compete with other economic priorities. Advance research, at least compared to model R&D, testing, and production, is relatively cheap and can produce useful spin-offs for the civilian sector. However, the costs of most programs would likely rise significantly the closer they move toward production. Therefore, China's defense industries can afford to do theoretical evaluation on a variety of systems and technologies, but will be forced to selectively choose systems for further development.

There is a widespread perception that the technological level of the Chinese defense industrial complex, to include quality control, is just too far behind to ever produce weaponry which could challenge U.S. supremacy. However, most within the PLA believe matching the technological level of the United States is not necessary to complicate U.S. power projection in the Asia-Pacific region. The predominant view within China, and that which was advocated by Deng Xiaoping, is that selective emphasis into a few "pockets of excellence" is enough.

Since the demise of the USSR and dissolution of any viable international export control regime, some of the world's most advanced technology has flooded into China. It is commonly argued that the rate of technology introduction is simply more than the defense industries can handle. China has had difficulties translating theory and design success into reliable weapon systems. Beijing is painfully aware of this shortcoming, and has taken steps to improve manufacturing through the widespread introduction of computer aided manufacturing, highly precise machine tooling equipment, and added emphasis on integrating research and production.

China's defense industries have been purchasing a broad range of foreign high tech equipment without much thought as to how to integrate various components into a system. Examples include their air traffic control and national telecommunications infrastructure modernization. Aware of their shortcomings, the defense industrial complex is seeking out international systems integration consultants. Security considerations and bureaucratic in-fighting, however, continue to slow progress.

Strategic Modernization and the Cross-Strait Military Balance.

Despite these obstacles, it is dangerous to underestimate China's ability to make significant leaps in capability over the next decade or two. Mistakes in assessing PLA modernization could be extremely costly to U.S. regional interests and even in U.S. lives. Nowhere is this more true than assessing the balance in the Taiwan Strait and the effect U.S. intervention could have.

China's strategic modernization has the potential to significantly alter the military equation in Taiwan Strait early in the next century. Emphasis on preemptive, long-range precision strikes, information dominance, command and control warfare, and integrated air defense could enable the PLA to defang Taiwan's ability to conduct military operations. While Beijing has a range of other options available, such as seizing an off-shore island or conducting missile exercises similar to those of March 1996, paralysis of the Taiwan military can allow Beijing to achieve its objectives with relatively few casualties and without widespread destruction of the island.

The following discussion provides one scenario which could bring into play the doctrine and strategy under discussion as well as systems under development. In 2010, for example, should the CMC decide to settle the Taiwan question once and for all, doctrinal writings and systems under development indicate the PLA's first objective could be to conduct parallel operations against Taiwan's early warning radar sites, SIGINT facilities, power plants, national-level command centers, and communications nodes in order to paralyze Taiwan's military command and control apparatus. PLA operations would require coordinated use of electronic warfare assets, anti-radiation missiles, land attack cruise missiles, ballistic missiles, special forces, and conventional air strikes. Preparation and coordination for such a complex operation would have to go undetected by Taiwan and U.S. intelligence. This would require strict emission security and passive counterspace measures, such as transmission of satellite warning messages, concealment, camouflage, and deception.

PLA electronic warfare operations would be directed against Taiwan's approximately 25 early warning radar sites. Through the GSD Fourth Department ELINT collection, the PLA likely has a library of the operating parameters of these radars, which can be used to program ship and airborne radar jammers. Without sophisticated ECCM, the performance of Taiwan's early warning radar network would likely be degraded. If radars are still operating, air-launched antiradiation missile strikes could either destroy the radars or at least force the operators to turn them off. Low flying PLA attack helicopters, if fielded by 2010, could be employed as well. Special operations forces could not only strike selected radar sites, but could also cut communications links connecting the radar sites to air defense command centers. The PLA would also likely attempt to jam, or if available by 2010, use long-range antiradiation air-to-air missiles to target Taiwan's E-2T AEW platforms should they be airborne.

Taiwan's surface-to-air missile sites, especially Modified Air Defense System (MADS) and Tien-kung I and II, would be also primary targets. Air defense radars and tactical C^2 links would be the focus of jamming operations, antiradiation, land attack cruise, and ballistic missile strikes. Like the early warning radars, the threat of antiradiation missile strikes could force operators to shut them down.

PLA doctrinal writings on command and control warfare strongly indicate they would target strategic command, control, communications, and intelligence centers as well. Major SIGINT facilities which would be monitoring some PLA activities would be targeted with ballistic and cruise missile strikes. Special forces could target antennas and cut communications links to command centers. Ministry of National Defense and Taiwan Air Force (TAF) command centers would also likely be attacked. Hardened command centers, however, are not easy to destroy. PLA operations, therefore, would likely be directed against the communications links, to include fiber-optic cables, connecting the command centers to the outside world. At the same time, PLA information operations would attempt to shut down Taiwan's military and civilian power and transportation computer systems through the insertion of viruses.

In parallel with strikes against Taiwan's C^4I infrastructure, the PLA Second Artillery may target Taiwan's eight primary airfields. Ballistic and cruise missile strikes against airfields would likely focus on runways, POL facilities, and barracks housing pilots. Airfields hosting E-2T, F-16s, and any electronic combat aircraft would likely take priority. If their missiles are indeed able to achieve an accuracy of 30-45 meters or better, the PLA would likely attempt to target specific taxiways as well as entrances to any underground aircraft storage facilities.

Should the Taiwan Air Force (TAF) be able to get aircraft off the ground, PLA air defense modernization and CCD programs could severely limit successful counterair or counterforce operations. One option for Taiwan to counter the ballistic and high performance cruise missile threat is to carry out offensive missions against the PLA bases, ships, and submarines which are carrying out the strikes. However, even if able to generate sorties in between missile or air strikes, finding and then reaching mobile Second Artillery launchers, missile garrisons, and command and control facilities in a dense air defense environment could be a daunting task.

The cumulative effect of simultaneous strikes against critical nodes in Taiwan's C^4I infrastructure, SAM sites, and airfields could be systemic paralysis. A temporary breakdown of Taiwan's early warning apparatus, command and control structure, major SAM sites, and airfields would allow a brief window of opportunity for the conduct of conventional air strikes against the same set of targets, as well as a broader range of targets. Ballistic and cruise missile strikes probably would not be able to completely shut down an airfield. However, missile strikes, combined with an information vacuum caused by heavy jamming and disruption of communications, could be enough to seriously complicate the TAF's ability to generate sorties.

This, in theory, provides a brief window for the PLAAF to follow up with second wave of strikes, this time using PLAAF assets with precision guided bombs. Through the use of ballistic, land attack cruise, and antiradiation missiles, and force multipliers such as electronic warfare and computer attacks, the PLAAF can offset any disadvantages it has in comparision to the TAF. The end result of these first two waves—the first using a combination of jamming, antiradiation, cruise, and ballistic missiles, and special operations forces and the second employing conventional air strikes—would be enough to effectively suppress Taiwan's information infrastructure and air defense system. Some Beijing-based analysts believe air superiority over Taiwan could be acheived in 45 minutes or less.[1]

Owning the skies over Taiwan, the PLA would have numerous options available. Beijing could simply establish a no-fly-zone in order to pressure Taiwan authorities to assent to their demands. If further hostilities are necessary, the PLA could leverage its air superiority to then achieve dominance of the waters around Taiwan. If occupying the island is necessary, a final step, after air and naval superiority are established, would be to transport PLA ground forces across the Strait. Owning the skies above and the waters around Taiwan, PLAAF and PLA(N) assets would provide the necessary firepower to ensure Taiwan ground force resistance would be minimal. PLA preparation of the battlespace through long-range precision strikes and information operations threatens Taiwan's ability to command and control its forces, and significantly reduces its costs of achieving air superiority—even with the PLA's backward air force. In short, emphasis on information dominance, missile forces, and air defense could decisively tip the cross-Strait military balance in Beijing's favor.

PLA Strategic Modernization and the United States.

The PLA's strategic modernization has implications for the United States. An increasingly capable PLA does not necessarily mean China is an emerging threat. Chinese strategy is based on developing a military that is relatively cheap, does not require a large infrastructure to maintain, but that nevertheless could complicate U.S. power projection ability should interests collide. There are signs that some within China believe that armed conflict with the United States is inevitable in the long term, especially over Taiwan. In the short to mid term, though, China must maintain a peaceful environment—and good relations with the United States—in order to develop its economy, S&T base, and military force.[2] While there is modernization underway across the board, only certain areas, such as C^4I and missiles, have been granted special attention. Successful fielding of only a handful of key systems could significantly hamper U.S. operations in the region.

The Taiwan scenario discussed above did not include U.S. involvement. One could, however, be devised. Open source writings and R&D efforts indicate that PLA planners do consider the possibility of U.S. intervention. In the scenario above, the PLA would conduct spasmodic, preemptive strikes to quickly establish information dominance and air superiority over the skies of Taiwan. U.S. aircraft carrier battle groups and USAF assets operating out of Okinawa, Japan and Guam may attempt to prevent or remove the PLA's air superiority.

China is devoting considerable resources toward preparing for potential conflict with the United States, especially over Taiwan. Despite its overwhelming victory in the Gulf War, Chinese analysts have concluded the U.S. military has vulnerabilities which can be exploited. For example, the United States continues to rely on a few fixed bases from which to project power into the region in a contingency situation. Heavy concentrations of aircraft flowing into bases will become increasingly easy to identify, as will aircraft carrier battle groups operating within the first island chain. C^4I assets, to include critical nodes in space, would be monitored as well.

The PLA is placing a premium on denying the United States and other potential adversaries information dominance. Fiber-optics, communications and operational security practices, and increasingly sophisticated CCD measures will make it difficult to correctly identify and target operational centers of gravity. In the unfortunate event of a major war, Beijing's investment into a highly capable air defense system, with emphasis on counterstealth, could significantly raise the human and material costs of attacking targets on the mainland, even if critical nodes can be identified. U.S. logistical bases and prepo facilities will become more vulnerable as well.

China's defense industrial establishment is investing in the development of components of a sensor network which, within the the next 15 years, could enable the PLA to detect, identify, and track ships operating around its periphery. Successful deployment of electronic reconnaissance and radar satellites, in particular, could have significant implications. ELINT and radar satellites could work well together to detect, identify, and track U.S. carrier battle groups and other vessels of interest. The combination is effective because they would be complementary and redundant. If the ship being tracked goes silent, the passive electronic reconnaissance satellite will not be able to pick it up but the SAR satellite will still register its position. If the ship attempts to jam the radar, the ELINT platform can easily pick it up. Using the spacecraft in combination, therefore, makes it extremely difficult for the enemy ships to hide, especially if linked to PLA ground-based SIGINT, surface wave over-the-horizon radars, submarines, and ostensibly civilian sentry vessels operating in the open ocean.

Chinese publications are already assessing that the United States could not sustain combat operations in defense of Taiwan. Analysts are carefully evaluating vulnerabilities in U.S. aircraft carrier battle groups. Shortcomings of aircraft carriers, some have concluded, include their prominent signature, tendency for bad weather to degrade operations, reliance on complex logistics, poor ASW and anti-mine capability, and difficulties in rapidly repairing damaged flight decks. Ways to counter aircraft carriers include mine laying, submarine operations, electronic warfare, attacking carrier-borne AEW assets, employment of large numbers of UAVs, and night operations. Aircraft carrier battle groups, say Beijing-based publications, are too reliant on highly vulnerable logistics ships and disruption of supply lines will force withdrawal of the carriers.

The PLA does not limit vulnerability assessments to the U.S. Navy. Assuming Japanese permission is granted, USAF operations out of Okinawa's Kadena Airbase, Chinese writers say, would be hampered by the distance beween Okinawa and Taiwan. Deploying USAF assets to Taiwan would open them up to attack by highly accurate ballistic and cruise missiles. Resupplying Taiwan by air or ship would also be vulnerable to attack. Analysts have also examined employment of ballistic missiles against carriers.[3] In short, China's force modernization, to a certain degree, appears to be geared toward countering U.S. ability to intervene in a cross-Strait conflict. This anti-access strategy is centered on targeting operational centers of gravity, including C^2 centers, airbases, and aircraft carrier battle groups located around the periphery of China.

China and the Revolution in Military Affairs.

Chinese open source writings often discuss information dominance, long-range precision strike, and asymmetrical warfare within the context of a revolution in military affairs (RMA) with Chinese characteristics. In fact, Chinese analysts have pointed out that the RMA could favor a country like China more than the United States. Ironically, one reason is the U.S. experience in the Gulf War conflict:

> It is the U.S. victory in the Gulf War that can hinder the further development of the U.S. military . . . the U.S. military has already become excessively reluctant to part from its existing military power and concepts—it is very possible future enemies will use this weakness and in the new RMA put forth more advanced thinking than the US.[4]

Another reason is the proliferation of information technology. Chinese observers believe that U.S. superiority in technology associated with information warfare is rapidly fading. To a great extent, some of the world's most advanced information technology is widely available on the commercial market. Strategists point out that the United States overestimates its ability to gain information superiority in the face of a determined adversary that has a well-established program to deny U.S. sensors its intelligence objectives.[5]

The RMA's highly precise weaponry, efficient array of sensors, and modern, survivable telecommunications infrastructure may be more conducive to denying the U.S. power projection than in making power projection the easy peace action which many military leaders expect. Little RMA technology gives the United States the capability to insert decisive force half-way around the world without large bases and platforms. The RMA shows no signs of reducing U.S. dependence on a few fixed air bases, ports, and carriers located close to the battlefield. Simply put, projecting power will be far more difficult than being a regional defender.

China, operating from its own territory, is growing in its ability to disperse assets, set decoys, confuse sensors, and distribute supplies well in advance. These technical changes will make offensive military action much more difficult. A well-prepared regional power like China can hide and disperse its forces much better than the United States. It is clear that the RMA, which stems from precise weapons, powerful sensors, and increased communications capability, could shift the balance of power to a country like China, despite clear U.S. superiority and what most analysts would still call a backward PLA.

Asymmetrical advantages which the PLA may acquire over time have political implications. As it becomes increasingly clear that power projection will become more costly, both in terms of lives and material, the United States may find that interests it previously judged vital will no longer warrant intervention. Thus, China may well defeat U.S. strategy without actual combat, the essence of Sun Tzu. It is no coincidence that as China develops an ability to complicate power projection, it is also beginning to use political rhetoric against the U.S. system of bilateral alliances and the maintenance of U.S. forward deployed forces in the region.

In the long term, China may hope to challenge unquestioned U.S. supremacy in space. Cooperation in space technology with former Soviet states, France, and Germany increases the chances of Chinese success in this area. China's rise in space comes at a time when Russia's space program is shriveling up due to a lack of funding. At one time, Russian leaders considered abandoning their manned space program and stated that orbiting Russian communication satellites may cease to exist at any time. The Russian space program has a severe shortage of launch vehicles as well.[6]

There are no easy solutions to potential challenges posed by PLA modernization. The first problem lies in analytical approaches to evaluating PLA modernization. Most within the U.S.-based PLA watching community are skeptical of

China's ability to modernize its military. One helpful measure would be a shift from concentration on the PLA's many shortcomings to a careful, sober examination of what the PLA could do now and in the future. The first step is recognizing that the PLA intentionally masks its capabilities as a fundamental approach to deterrence. While the U.S. purposefully displays its capabilities to effect deterrence, China hides its strengths and weaknesses as a means to inject uncertainty into the minds of potential adversaries.

Greater effort must be made to convince China that being a cooperative and transparent member of the international community is in China's interests. We must continue to expand our network of contacts with the PLA in order to better understand their long-range objectives and guide them toward responsible behavior within the region. In our relationship with the PLA, though, we must be realistic in anticipating how far the PLA will go in opening their doors to outside observers. Transparency is anathema to their doctrinal emphasis on concealment and deception. Nevertheless, maintaining a wide ranging dialogue will aid in building confidence and reducing suspicions. In short, a PLA which is able to complicate U.S. power projection through asymmetrical means does not automatically mean China is a threat to the United States if the relationship is well-managed.

ENDNOTES - CHAPTER 6

1. For an assessment of Taiwan vulnerabilities and PLA attack strategies using combination of ballistic and cruise missiles, electronic warfare, and aircraft, see Yuan Lin, "The Taiwan Strait is No Longer a Barrier—PLA Strategies for Attacking Taiwan," *Kuang Chiao Ching (Wide Angle)*, April 16, 1996, No. 283, pp. 14-19. *Wide Angle* is a Hong Kong-based publication with close links to the PLA. For an in-depth mainland discussion on use of ballistic missiles against airfields, see Li Xinyi, "On the Air Supremacy and Air Defense of Taiwan and China: Is Taiwan an 'Unsinkable Aircraft Carrier'?," *Taiwan de Junbei (Taiwan Military Preparations)*, July 1, 1996, pp. 11-18, in *FBIS-CHI*-97-323. Li is the analyst who asserts the Taiwan Air Force would be brought to its knees in about 45 minutes.

2. See, for example, Li Tzu-ching, "CPC Thinks China and the United States Will Eventually Go To War," *Cheng Ming*, May 1997, pp. 15-16, in *FBIS-CHI*-97-126.

3. Ying Nan, "Hangmu de Biduan Ji Fanhangmu Zuozhan" ("Shortcomings of Aircraft Carriers and Anti-Carrier Operations"), *Xiandai Junshi*, January 1998, pp. 13-15; Zhu Bao, "Discussion of Technical Means to Attack Aircraft Carriers With Tactical Ballistic Missiles," *Hubei Hangtian Keji*, 1997 (1), pp. 46-49, in *CAMA*, 1997, Vol. 4, No. 4. Zhu is from the 066 Base Institute of Precision Machinery. Employment of ballistic missiles against aircraft carriers could be a reality if the First Academy is able to master maneuvering re-entry vehicles and terminal guidance technologies. Also see Chang Lan, "Analysis of the Defense System of American Aircraft Carriers," paper presented during November 1997 conference of National Missile Designers Network, in *CAMA*, Vol. 5, No. 3. Chang is from the Beijing Institute of Space Systems Engineering, CASC's primary missile systems design institute. In addition, see "U.S. Military Intervention in Cross-Strait Conflict Seen As Unlikely," *Taiwan de Junbei*, July 1, 1996, pp. 76-79, in *FBIS-CHI*-97-302; and Su Qi, "Intervention in Taiwan Question Seen As Harmful to U.S. Interests," *Taiwan de Junbei*, July 1, 1996, pp. 72-75, in *FBIS-CHI*-97-302.

4. Zhu Xiaoli and Zhao Xiaozhuo, *Mei'E Xin Junshi Geming* (*The United States and Russia in the New Military Revolution*), Beijing: AMS Press, 1996, pp. 40-45.

5. *Ibid.*

6. "Russia Official Says Finances Threaten Nation's Space Aims," *Boston Globe*, December 20, 1996.

APPENDIX I

CHINA AEROSPACE CORPORATION ORGANIZATION

Director: Liu Jiyuan. Russian educated, former vice-president of CALT.

Vice-Directors:

- Wang Liheng. Former Third Academy vice-president.
- Luan Enjie. Navigation and control expert.
- Bai Bai'er. Educated at Harbin Institute of Technology.
- Xia Guohong. PhD from University of California.

Total Number of CASC Employees: 270,000

Direct Reporting Elements:

- General Office. Manages daily duties of CASC headquarters.
- Comprehensive Planning Department. Develops long-range developmental strategies and policies.
- Scientific Research and Production Department. Manages fiscal year planning for commercial space launches and military and civil production.
- S&T Department. Oversees basic research projects.
- International Cooperation Department. Responsible for international exchanges and negotiations.

- Security Department. Safeguards information on China's space and missile R&D and production.

- China Great Wall Industry Corporation (CGWIC). Markets a wide range of items, to include satellite launch services and a broad range of civil and military products. China Precision Machinery Import/Export Corporation (CPMIEC), a subordinate entity under CGWIC, has dominated the missile sales realm.

701st Research Institute	Beijing Institute of Aerodynamics. Conducts windtunnel testing for CASC systems.
707th Research Institute	Institute for Astronautics Information. Collects, analyzes, and distributes information for use throughout the aerospace community.
708th Research Institute	Institute of Space Standardization
710th Research Institute	Institute of Computer Systems
307 Factory	Nanjing Chenguang Machine Factory. Final assembly for solid fueled missile systems. Employs 7,800 people.

- Science and Technology Committee.

- China Resource Satellite Application Center. Directed by Wu Meirong. Conducts liaison with others in remote sensing community.

- Beijing Simulation Center. Asia's largest simulation facility.

- Beijing Space Technology Test Center

- Shenyang Xinguang Dynamic Machinery Company
- Shenyang Xinle Precision Machinery Company
- Xinyang Company

1st ACADEMY

CHINA ACADEMY OF LAUNCH TECHNOLOGY

Director: Shen Xinshun

Vice-Director: Xu Dazhe

Location: Nanyuan, adjacent to PLAAF Nanyuan Airfield in southern suburbs of Beijing.

Mission: R&D and production of launch vehicles; liquid fueled surface-to-surface missiles; and solid-fueled surface-to-surface and submarine launched missiles. Employs over 27,000 personnel. Business name of Beijing Wanyuan Industry Corporation. Organized into 13 research institutes and seven factories.

Important departments, institutes, and factories:

1st Planning Department	Beijing Institute of Astronautical Systems Engineering (Liquid Systems)
4th Planning Department	Beijing Institute of Electro-mechanical Systems Engineering (Solid systems engineering)
11th Research Institute	Beijing Institute of Liquid Rocket Engines. Also known as Beijing Fengyuan Machinery Company. Directed by Liu Guoqiu and employs over 900 people. Operates 067 Base liaison office and test site in southwest suburbs of Beijing.

12th Research Institute	Beijing Institute of Automatic Control. Colocated with 2nd Academy facilities on Yongding Road in western Beijing. Over 800 personnel assigned. Established in 1958. Engaged in R&D of missile related guidance technology to include GPS exploitation. Directed by Dong Ruohuan.
13th Research Institute	Beijing Institute of Control Devices. R&D of inertial instrument technology such as gyros and accelerometers. Over 700 assigned. Ding Henggao served as deputy director. Current director is Sun Zhaorong.
14th Research Institute	Beijing Special Electromechanical Institute. Warhead development. Closely associated with CAEP in effort to miniaturize warheads. Employs 800 personnel and directed by Wu Zhaozong.
15th Research Institute	Beijing Institute of Special Engineering Machinery. Ground equipment, to include launch control and missile launcher survivability. Directed by Bao Yuanji.
702nd Research Institute	Beijing Institute of Structure and Environmental Engineering. Directed by Yang Yongxin.
703rd Research Institute	Beijing Research Institute of Materials and Technology. Directed by Mao Huamin.

704th Research Institute	Beijing Research Institute of Telemetry (BRIT). Since 1991, engaged in exploitation of GPS. Employs over 1000 personnel. Li Bingchang.
200 Factory	Guanghua Radio Factory. Control system electronic components.
210 Factory	Beijing Jianhua Electronic Instrument Factory. Inertial components. More than 1200 employees. Close association with 13th Research Institute. Located in Nanyuan.
211 Factory	Capital Space Machinery Corporation. General Assembly Plant (Liquid systems). Located in Nanyuan complex.
230 Factory	Beijing Xinghua Machinery Factory. Located on Yongding Road.
7107 Factory	Inertial Devices Factory. Located in Baoji. Associated with 230 Factory.
Beijing Experimental Factory	Electronic hydrolic servo systems. Located in Muxidi, West Beijing.
Beijing Wanyuan Sealing Factory	

2nd ACADEMY

CHANGFENG ELECTROMECHANICAL TECHNOLOGY DESIGN ACADEMY

Director: Chen Dingchang. Replaced Liu Congjun.

Location: Yongding Road, western suburbs of Beijing

Mission: R&D and production of air and missile defense, ASAT, and associated radar systems.

Number of personnel: 12,800

Important departments, institutes, and factories:

2nd Planning Department	Beijing Institute of Electronic Systems Engineering. Department-level institute responsible for air/missile defense and ASAT systems engineering. Employs over 700 engineers. Directed by Yuan Qi.
17th Research Institute	Beijing Institute of Control and Electronic Technology. Located in Muxidi, West Beijing.
23rd Research Institute	Beijing Institute of Radio Measurement. R&D of radar systems. More than 1300 people. Located on Yongding Road and directed by Huang Huai.
25th Research Institute	Beijing Institute of Remote Sensing Equipment. Radar and optical terminal guidance systems. Directed by Sun Zhaoxin.
203rd Research Institute	Beijing Institute of Radio Metrology and Measurement. More than 250 engineers. Directed by Miao Fuquan.
204th Research Institute	Beijing Institute of Computer Applications and Simulation Technology. Computer software and simulation technology.

206th Research Institute	Beijing Institute of Mechanical Equipment (Jixie Shebei). Launchers and other ground equipment. Employs 570 people. Directed by Qin Ye.
207th Research Institute	Beijing Institute of Environmental Features. R&D into target characteristics. Microwave, optical, and laser environmental engineering. Manages microwave anechoic chamber and laser laboratory. Space observation. Directed by Zhao Ji.
208th Research Institute	Information Center and publisher of 2nd Academy's journal, *Systems Engineering and Electronics Technology*.
210th Research Institute	Xian Changfeng Electromechanical Institute. Electromechanical systems engineering. Employs over 1500 personnel. R&D on telemetry, dynamic strength and heat/cold testing. Directed by Huang Wangsheng.
706th Research Institute	Computer development
112 Factory	Xinfeng Machinery Factory. Final assembly plant for SAMs, etc.
123 Factory	Air Defense Missile Warheads
283 Factory	Ground Control Systems
284 Factory	Control Systems Manufacturing. Xinjian Power Machinery Plant. Located on Yongding Road.

786 Factory   Radar and Guidance Systems
                           Plant

3rd ACADEMY

HAIYING ACADEMY OF ELECTROMECHANICAL TECHNOLOGY

Director: Wang Jianmin

Location: Yungang, in southwestern suburbs of Beijing

Number of personnel: 14,500 in 10 research institutes and two factories.

Mission: R&D and production of anti-ship and land attack cruise missiles and associated systems. Established in 1961. S&T Commission headed by Yao Shaofu.

3rd Design Department	Beijing Institute of Electromechanical Engineering. Responsible for anti-ship/land attack cruise missile design and systems engineering. Directed by Li Huiting. Over 1200 personnel assigned.
31st Research Institute	Power Machinery Research Institute. Develops cruise missile propulsion systems. Employs more than 1,300 personnel. Established in 1957, predating establishment of 3rd Academy. Directed by Zhang Zhenjia.
33rd Research Institute	Beijing Institute of Automated Control Equipment. Cruise missile autopilot and inertial naval systems. Located in Yungang.

35th Research Institute	Huahang Institute of Radio Measurement. Located in Hepingli, Beijing.
310th Research Institute	Information collection, analysis, and dissemination. Produces advocacy papers on cruise missile systems.
8357th Research Institute	Jinhang Institute of Computing Technology. Control systems and on-board computer systems. Located in Tianjin. 470 personnel.
8358th Research Institute	Jinhang Institute of Technical guidance. Located in Tianjin. Publishes technical journal *Infrared and Laser Engineering*.
8359th Research Institute	Beijing Special Machinery (Tezhong Jixie) Institute. Cruise missile launching equipment (tube, air, etc). Russian exchanges.
119 Factory	Autopilot systems
159 Factory	Xinghang Electromechanical Factory. Located in Yungang.
239 Factory	Beijing Hangxing Machine Building Factory. Located on Hepingli Street in Beijing. General assembly of cruise missiles. Led by Hu Zongyin, over 3000 assigned.
558 Factory	Autopilot and altimeter production.
781 Factory	Terminal Guidance System Plant
786 Factory	Ground Tracking Radar Factory

5013 Factory   Warhead Plant

4th ACADEMY

Director: Ye Dingyou

Number of personnel: 3500

Mission: R&D and production of solid fueled motors for ballistic missiles and satellite kick motors. Founded in 1965. Fourth Academy corporate name of Hexi Chemical Machinery Company.

41st Research Institute	Shaanxi Institute of Power Machinery. Solid rocket motor design. Located near Hohhot. Directed by Wang Desheng.
42nd Research Institute	Red Star Chemical Institute of Hubei
43th Research Institute Non-Metallic	Shaanxi Institute of Materials and Technology. Filament winding machines.
44th Research Institute	Shaanxi Institute of Electronics
46th Research Institute	Hexi
47th Research Institute	Xiangyang Chemical Machinery Corporation
7414 Factory	Shaanxi Hongchuan Machinery Factory
7416 Factory	Shaanxi Changhong Chemical Plant. Assembly factory.
7422 Factory	Xian Space Lanling Factory
7424 Factory	

Shaanxi Xianfeng Institute of Machinery

Hexi Corporation

- Synthetic Chemical Engineering Institute
- Inner Mongolia Hongguang Machinery Plant
- Inner Mongolia Hongxia Chemical Plant
- Inner Mongolia Power Machinery Plant

5th ACADEMY

CHINA ACADEMY OF SPACE TECHNOLOGY

Director: Xu Fuxiang (recently replaced Qi Faren after series of satellite failures)

Vice-Directors: Zhang Guofu, Yuan Jiajun, and Ma Xingrui

Location: Haidian district of northeastern Beijing, on Baishiqiao Road.

Mission: R&D and production of communications, space-based ISR systems, and weather satellites. Laying groundwork for future navigation satellites; data relay satellites; space shuttle; and space station. Established 1968. With 10,000 personnel, oversees 14 research institutes and factories. Current capability to produce 4-6 satellites per year.

501st Research Institute	Beijing Institute of Spacecraft Systems Engineering (Department-level entity). Established in 1968. Responsible for satellite systems engineering.
502nd Research Institute	Beijing Institute of Control Engineering. R&D of attitude control systems. Established

	1956 under CAS. Employs over 1400.
503rd Research Institute	Beijing Institute of Satellite Information Engineering. R&D of satellite applications and communications technology, including ground segments of FY-2 weather satellite and GPS receivers. Involved in China's indigenous effort to develop satellite nagivation system (RDSS). Established 1986. Employs more than 300 personnel.
504th Research Institute	Xian Institute of Space Radio Engineering. R&D of space communications, remote sensing, and spacecraft TT&C. Produces space electronic systems, including TWTs, CCD camera and microwave data links, and antennas. Personnel total 1200.
508th Research Institute	Beijing Institute of Space Machinery and Electronic Engineering. Located adjacent to 1st Academy facilities in Nanyuan. R&D of remote sensing and recoverable vehicle technology.
510th Research Institute	Lanzhou Institute of Physics. Conducts research into optical cryogenics, microgravity, and radiation effects.
511th Research Institute	Beijing Institute of Environmental Test Engineering. Facilities in Beijing and Huairou.

513rd Research Institute Yantai Telemetry Technology Institute

529 Factory Beijing Orient Scientific InstrumentFactory. Final assembly for satellite systems. 1200 personnel.

8th ACADEMY

SHANGHAI ACADEMY OF SPACE TECHNOLOGY

Director: Zhang Wenzhong

Mission: Created in 1961, employs 30,000 personnel in 17 institutes and 11 factories. SAST supplies the first two stages of the LM-2, LM-3, and LM-4 and Fengyun meteorological satellites.

- Shanghai Institute of Electromechanical Engineering (8th Design Department). Directed by Jin Zhuanglong. Employs 478 personnel.

- Shanghai Institute of Satellite Engineering (509th Research Institute). SAST's key unit for satellite systems engineering and environmental testing. Established in 1969 and employs more than 600 people. Directed by Lu Zili. Primary products include FY-1 and FY-2 satellites. (CAST brochure has this institute under CAST)

- Shanghai Precision Machinery Research Institute

- Shanghai Institute of Power Machinery

- Shanghai Institute of Radio Equipment (802nd Institute). SAM guidance and fuzes.

- Shanghai Institute of Precision Instruments. Guidance systems. Xinyue Institute.

- Shanghai Institute of Electronic Communications Equipment Engineering

- Shanghai Institute of Electromechanical Equipment

- Shanghai Xinfeng Chemical Engineering Institute. Propellant technology.

- Shanghai Institute of S&T Information for Electromechanical Engineering

- Shanghai Institute of Precision Metrology and Test Engineering

- Shanghai Xinwei Electronic Equipment Research Institute (809th Institute). LV and tactical weapons computer automation launch control systems design and satellite control computers. Established 1979.

- Shanghai Xinli Institute of Power Equipment. Engines and motors.

- Shanghai Institute of Space Power Sources

- Shanghai Spaceflight Automatic Control Equipment Research Institute

- Shanghai Institute of Spaceflight Telemetry, Control, & Telecommunications Engineering

- Shanghai Institute of Video & Telecommunications Equipment Engineering

- Shanghai Spaceflight Architecture Design Institute

- Shanghai Xinzhonghua Machinery Factory

- Shanghai Xinjiang Machinery Factory

- Shanghai Xinxin Machinery Factory

- Shanghai Xinhua Radio Factory
- Shanghai Xinya Radio Factory
- Shanghai Xinguang Telecommunications Factory
- Shanghai Xinyu Power Supply Factory
- Shanghai Xinli Machinery Factory
- Shanghai Broadcast Equipment Factory
- Shanghai Instrument Factory
- Shanghai Wire Communication Factory

9th ACADEMY

CHINA ACADEMY OF SPACE ELECTRONICS TECHNOLOGY

Director: Tao Jiaqu

Location: Headquartered in Beijing's Haidian district (8 Fucheng Road), sites in Nanyuan and near Xian. Established in 1993.

Mission: Development of specialized computers, integrated circuits, and other microelectronic devices in support of CASC projects.

Number of employees: More than 10,000 in 9 institutes, 10 manufacturing plants, and 5 technical centers.

Important institutes and factories:

771st Research Institute	Lishan Microelectronics Institute. Established 1960s. R&D and manufacturing of missile satellite-related computers and

integrated circuits. 3800 personnel.

165 Factory

061 BASE

CHINA JIANGNAN SPACE INDUSTRY GROUP

Director: Li Guozhong

Mission: Development of systems associated with surface-to-air missiles.

Scope: 35 institutes, factories, and companies

Number of personnel: 6400

Location: Zunyi and Kaishan, Guizhou Province

- 302nd Research Institute (General Institute of Military Products)
- Jiangnan Electromechanical Design Institute
- 38th Research Institute
- 303rd Research Institute
- Wujiang Machinery Factory
- Nanfeng Factory
- Xinfeng Instrument Manufacturing Corporation. Tracking and control systems.
- Qunjian Machinery Factory
- Chaohui Electromechanical Factory
- Meiling Factory
- Honggang Electromechanical Factory

- Guizhou Gaoyuan Machinery Factory. SAM launchers.

062 BASE

SICHUAN AEROSPACE CORPORATION

Director: Yu Ruihua

Number of Personnel: More than 20,000

Headquarters: Chengdu, Sichuan province

Mission: Development of systems associated with liquid-fueled ballistic missiles, launch vehicles, and anti-ship missiles.

- Chongqing Aerospace Electromechanical Design Institute (800 personnel)

- Sichuan Changzheng Mechanical Factory. Located in Wanyuan, in northern Sichuan (5,000 personnel).

- Chongqing Bashan Instrument Factory. Telemetry equipment.

- Fenghuo Machinery Factory. Servo-mechanical devices.

- Liaoyuan Radio Factory. Space flight controls. Located in Xuanhua, Sichuan province.

- Tongjiang Machinery Factory. Metals processing.

- Mingjiang Machinery Factory. Located in Dachuan.

- Pingjiang Instrument Factory. Control systems. Located in Dachuan.

- Chuannan Machinery Factory. Missile system ignitors.

O66 BASE

SANJIANG SPACE GROUP

Director: Cao Lijia

Number of personnel: 17,000.

Headquarters: Xiaogan, north of Wuhan. Production centered in Yuan'an, in western Hubei province. Maintains office in Wuhan.

Mission: R&D of solid-fueled tactical ballistic missiles and stealth/counterstealth technology. Base 066 established in August 1969 as Third Line production base for 3rd Academy anti-ship missiles. In 1975, began independent R&D of M-11 missile, designed by Wang Zhenhua, who completed system R&D in 1984.

- Sanjiang Space Group Design Institute

- Hubei Redstar (Hongxing) Chemical Institute, 42nd Research Institute, located in Xiangfan, Hubei province.

- Hubei Hongfeng Machinery Plant. Established 1970. Electromechanical integration. Located in Yuan'an.

- Wanshan Special Vehicle Machinery Factory. Located in Yuan'an.

- Hubei Jianghe Chemical Factory. Located in Yuan'an

- Xianfeng Machinery Factory. Located in Yuan'an.

- Wanli Radio Factory. Located in Yuan'an.

- Honglin Machinery Factory. Located in Xiaogan.

- Hubei Hongyang Machinery Factory. Located in Yuan'an.

- Jiangbei Machinery Factory. Located in Yuan'an.
- Wanfeng Factory. Located in Yuan'an.

067 BASE

SHAANXI LINGNAN MACHINERY CORPORATION

Director: Hu Hongfu

Mission: R&D base for liquid engines and inertial guidance systems for launch vehicles. Oversees five research institutes and four factories.

Number of personnel: More than 1,100

- Shaanxi Engine Design Institute
- Beijing Fengyuan Machinery Institute
- Shaanxi Institute of Power Test Technology
- Xian Changda Precision Electromachinery Institute
- Shaanxi Hongguang Machinery Factory
- Shaanxi Cangsong Machinery Factory
- 16th Research Institute
- 165 Research Institute
- 204 Factory
- 710 Factory
- 7103 Factory (Hongguang)
- 7107 Factory
- 7171 Factory. Inertial devices.

068 BASE

HUNAN SPACE AGENCY

Director: Hu Zhigang

Mission: Aerospace electromechanical equipment associated with surface-to-air missiles, and R&D of special materials. Oversees one research institute and five factories.

Location: Changsha, Hunan province

Important research institutes and factories:

- 7801 Research Institute. Directed by Yan Hailiang. Located in Changsha, Hunan province.

- 7803 Factory. Directed by Huang Yaoyue, produces superhard materials (chaoying cailiao). Hunan Taishan Machinery Factory.

- 804 Factory

- 861 Factory

- Hunan Zhujiang Instrument Factory

- Hunan Electromechanical Instrument Factory

YUNNAN SPACE GROUP

Director: Wang Shirong

Mission: Formerly tied to 3rd Academy. Established 1969. Formerly a Third Line unit within Yunnan province, moved to Kunming in 1987. Current responsibilities unclear. Manages six factories and eight companies.

Number of Personnel: 3,500

HARBIN INSTITUTE OF TECHNOLOGY

Academy of Astronautics

Academy of Material Science and Engineering

Department of Astronautics and Physics

Department of Applied Chemistry

Department of Applied Physics

Department of Communications Engineering

Department of Computer Science and Engineering

Department of Control Engineering

Department of Electrical Engineering

Department of Mathematics

Department of Mechanical Engineering

Department of Power Engineering

Department of Precision Instrumentation

Department of Radio Engineering

Department of Space and Opto-Electronic Engineering

Robotics Research Institute

Plating Research Center

Analysis and Measurement Center

Inertial Navigation Test Equipment Center

Simulation Center

Sources: *Zhongguo Hangtian (China Space News)* (weekly), July 1992-June 1995 and July 1996-January 1997; *China Today: Defense Science and Technology,* 1993; John Lewis and Xue Litai, *China's Strategic Seapower,* 1993; brochures from 1st, 2nd, 3rd, 4th, 5th, and 8th Academies, and various research institutes; *China Astronautics and Missilery Abstracts* (selected issues); John Lewis and Hua Di, *China's Ballistic Missile Development;* and *A Survey of China's Space S&T Industry*, 1994, Beijing: CASC.

APPENDIX II

THE LEGEND OF QIAN XUESEN

The father of China's space and missile industry is Qian Xuesen. Sent to the United States on scholarship in 1935, Qian was educated at Massachusetts Institute of Technology and CalTec and became one of the initial cadre of the Jet Propulsion Laboratory. As one of the world's foremost experts in propulsion and aerodynamics, Qian worked on a number of advanced aircraft and missile projects. In 1944, General Henry Arnold recruited Qian to work as a scientific consultant to the Army Air Corps' Scientific Advisory Group. Their task was to search nationwide and abroad for developments that would make American air power the very best in the world. Given an Air Force officer commission, Qian traveled throughout the country and drafted a report, *Future Trends of Development of Military Aircraft*, one of the U.S. Air Force's first long-range studies. At the conclusion of World War II, Qian and the Scientific Advisory Group traveled to Germany to interview German rocket scientists, including Werner von Braun.

Upon his return, Qian played a major role in the Scientific Advisory Group's production of *Toward New Horizons*, the blueprint for the long-range development of the U.S. Air Force. Qian led writing of aspects of the report which dealt with aerodynamics, pulsed engines, ramjet engines, solid and liquid rockets, and jet-propelled supersonic wing missiles. Between 1945 and 1950, as a member of the Scientific Advisory Board, he authored futuristic concepts including nuclear powered aircraft propulsion, manned space flight, and rocket powered transcontinental aerospace vehicles that travel at speeds in excess of 10,000 mph.

Qian Xuesen's promising career as a leading U.S. scientist and visionary was cut short, however, in the mania of the McCarthy era. Accused of being a communist, he was stripped of his security clearances. Under a cloud of suspicion, Qian was deported to China in 1955. In Beijing, Qian joined a group of other scientists and engineers who had returned to China after the establishment of the People's Republic of China (PRC) in 1949. Bent on extracting revenge on the United States, Qian quickly gained the confidence of Mao Zedong and Zhou Enlai and passionately accepted the responsibility of leading the development of China's aerospace capability. Qian was obsessed with the belief that China could do anything that the United States and the West could do, and he instilled this view in the cadre of engineers around him.

Qian's first task was to assemble a team of foreign-trained engineers and establish an aerospace research and development organization. In 1956 Qian shaped a plan that bears a striking resemblance to the U.S. Air Force's *Toward New Horizons*, which prompted China to adopt a long-range perspective in its weapons development. The plan placed emphasis on atomic energy, missiles, computer science, semiconductors, electronics, and automation technology. Qian also emphasized the exploitation of foreign—especially U.S.—technical materials to use as guides for indigenous development. Most importantly, Qian convinced the Chinese government that missile development should take precedence over aircraft development.

Based on a February 1956 proposal of Qian's, Zhou Enlai approved the establishment of the Fifth Academy of the Ministry of National Defense. Qian and the Fifth Academy lacked even the most fundamental resources, such as rubber, and tools to accomplish their objectives. In 1957, Qian led a military delegation to Moscow to lay the foundation for a Sino-Russian technical relationship. The relationship involved hundreds of Soviet engineers working in Chinese research institutes, Chinese students studying in the Union of Soviet Socialist Republics (USSR), and the transfer of technical designs to China. In all, the Soviets and the Chinese were engaged in 343 contracts and 257 technical projects. In August 1960, however, the relationship was abruptly ended.

During the next two decades, Qian guided the development of a family of ballistic and cruise missiles, aerodynamic testing facilities, satellites, and a TTC network, and was a strong advocate of nuclear power. Qian was harassed during the Cultural Revolution but was a crucial factor in the Central Committee's declaration of the space and missile industry as a top priority and therefore immune from the ravages of the Cultural Revolution.

In the years that followed, Qian became identified with the conservative factions of the government. In 1977 Qian spoke out against Deng Xiaoping and his supporters, including Commission of Science and Technology for National Defense Director Zhang Aiping. Qian's opposition to Deng and his dabbling in the supernatural resulted in a loss in credibility among China's scientific and technological elite.

Qian, however, regained status after supporting Deng Xiaoping's actions in cracking down on anti-government protests in 1989. Qian's protégé, Song Jian, was appointed as State Science and Technology (S&T) Committee chairman, and Qian again became active in scientific circles. Qian was showered with awards and accolades in 1991. Qian's most recent influence was felt during the March 1994 COSTIND S&T Committee Meeting when he pressed the Chinese leadership and COSTIND to adopt a visionary outlook and establish technologies and systems such as remote sensing satellites, hypersonic aerospace planes, adaptive optics, and communications systems as national priorities. Qian currently serves as special advisor to COSTIND's S&T Committee.

In sum, Qian's legacy still exists to this day. First of the Qian legacies is the confidence that China can match the West in technological development. The desire to develop and field systems simply for status sake is a driving force behind China's research and development strategy. Another legacy is Qian's preference of missiles over aircraft. In effect, he influenced the Chinese leadership to place strategic nuclear and missile industries first in the defense industrial hierarchy, including greater funding and prestige. Its historically lower status plagues the aviation industry to this day. Qian also instilled a forward looking approach to defense S&T strategy. He imbued a generation of China's scientific elite with the need to be forward looking and bold in their engineering ambitions.

APPENDIX III

SPACE SUPPORT FOR STRATEGIC MODERNIZATION

Development of a viable space force is a vital aspect of China's strategic modernization. There appears to be a clear long-range path to support an emerging doctrine which views space as an arena for competition. The basic building block of China's future capability in space is their launch vehicles. Up until the last few years, Chinese launch vehicles had one of the best success rates in the world, opening the door for international launch services. A recent spate of failures, however, has reduced foreign confidence. Domestic confidence has continued, though, and they are proceeding with other space transportation programs, to include a manned space capsule, a space shuttle, and preliminary research into single stage-to-orbit transatmospheric vehicles. China's premier space research and development entity, the China Academy of Space Technology (CAST) under CASC, is also developing a spectrum of new generation satellites.

China's Space Leading Group, under the State Council, oversees and coordinates all space activities in China. The group is composed of six members: COSTIND director, State Science and Technology Commission (SSTC) deputy director, vice-minister of the Ministry of Foreign Affairs (MFA), deputy director of the State Planning Commission, and the director of CASC. The Space Leading Group coordinates among various industries and agencies.[1]

China has several approaches to developing a space capability to meet the demands of 21st century warfare. First is development of smaller boosters able to launch satellites at a moment's notice in a contingency. Solid-fueled launch vehicles would probably be a natural transition after fielding the next generation of solid-fueled missiles around the turn of the century. Second is a movement toward various constellations of small satellites, which are more resistant to anti-satellite (ASAT) weapons and can be produced cheaply in large numbers. China is already investing in the U.S. Iridium program, a constellation of 66 small satellites, and is prepared to invest in their own indigenous answer to Iridium. China is also working with others in the region to jointly develop a smallsat constellation of imaging satellites. The next area is counterspace. Open source literature strongly suggests China's ASAT program is in the model development stage where the space industry is identifying various design proposals for seekers and propulsion systems.

CASC has a huge, bloated space infrastructure waiting to be filled to capacity. CAST's 529 Factory and SAST's 509 Factory currently only manufacture a few satellites a year. China Academy of Launch Technology's (CALT) 211 Factory only assembles a few launch vehicles, but, as of 1994, has the capacity to produce up to 10 vehicles a year. COSTIND's space tracking network provides tracking, telemetry and control (TT&C) services to a handful of domestic satellites.[2]

Space Transportation.

Essential to China's future in space is its launch vehicle infrastructure and systems. COSTIND has operated a network of space launch bases since their space program began in the 1970s. Launch centers include the following:

- Xichang Space Launch Center (XSLC). Xichang is China's primary site for the launch of space vehicles intended for geosynchronous orbit. The site is equipped with two launch pads, one for LM-3 launches and the other for LM-2E, LM-3A, and LM-3B vehicles.

- Jiuquan Space Launch Center. Jiuquan is used for the launch of remote sensing and scientific satellites, and will be center for China's manned space program. Primary LVs include LM-2C and LM-1D.

- Taiyuan Space Launch Center. Taiyuan is used for polar orbit satellites and plays a major role in launching the Iridium satellite constellation.

To augment COSTIND's three main launch centers, Chinese engineers are examining alternative launch modes for the 21st century. Concepts under consideration include a small solid-fueled launch vehicle which is air-launched; a sea-based launch platform; and a compressed-air launch system. These new approaches to space launches are viewed as simple, mobile, and capable of accommodating heavy loads. The lead COSTIND organization for exploring future launch modes is Beijing Academy of Special Engineering Design.[3]

There is a large volume of literature which outlines China's current family of launch vehicles, which includes several variants of the Long March (LM)-1, LM-2, LM-3, and LM-4. Their heaviest lift launch vehicle is the LM-3B which integrates a cryogenic third stage engine. All launch vehicles are manufactured by the CASC First Academy (CALT) and Eighth Academy in Shanghai. Current efforts are aimed at producing a small launch vehicle which can send small satellites into low and medium orbits. Critical to the small effort is microelectronics technology, new materials, and low power loss electronic devices.[4]

CALT is also striving to further increase the carrying capacity of their heavy launch vehicles. The launch vehicle which will carry China's manned space capsule will be a variant of the LM-2E, likely designated as the LM-2E(A). It will be launched from Jiuquan Space Launch Center and able to lift 12 tons to low earth orbit. CALT is looking at developing a nontoxic propellant launch vehicle with a single engine over 100 tons and a vehicle whose single cryogenic liquid oxygen/liquid hydrogen engine is capable of 50 tons of thrust.[5]

China is working to develop a new *Ariane V* class heavy-lift booster which utilizes cryogenic oxygen/hydrogen engine in the first stage. CALT currently utilizes a cryogenic third stage on the LM-3A and LM-3B launch vehicles. CALT is also researching large kerosene/oxygen engines, building upon engines purchased from Russia in 1993. China hopes to field a new generation launch vehicle which has the capacity to hurl 20 tons into low earth orbit. This vehicle will serve as the basis for China's lunar and planetary exploration.[6]

The Shanghai Academy of Space Technology's LM-4 will have the ability to launch dual payloads and is slated to boost the new FY-1C and an 880-pound magnetosphere research satellite in late 1998. Other modifications include a new third stage which will give the vehicle the capability to loft medium sized (2,200-3,300 pounds) payloads into highly elliptical orbits. The new stage will also have a restart capability.[7]

Launch Business.

China has made the launching of satellites a money making venture. Since the first commercial launch of ASIASAT 1, COSTIND and CASC have launched several more, each bringing in an average of $40-70 million. COSTIND and CASC are expected to increase the cost of launch services to bring them on a par with western launch providers. Recent launches include a Philippine satellite, Mabuhay, which was launched by a LM-3B, during summer 1997.[8]

Revenues from the commercial launch arena are significant. Since 1991, COSTIND and CASC have grossed approximately $500 million in U.S. dollars (10 launches, average 50 million each). With the cost of a launch vehicle estimated at five to ten million U.S. dollars, the commercial launch business is extremely profitable. In fact, one commercial launch matches the entire annual CASC investment into dual-use space technology. Revenues are shared between COSTIND and CASC, with the proportion varying based on the specific contract. Within CASC, profits are shared between CALT entities involved in manufacturing the launch vehicle; China Great Wall Industry Corporation, the marketing and contracting agent; and CASC headquarters.[9]

Despite a recent spate of failures, China's launch business will increase in the future. COSTIND and CASC have contracted to launch 12 U.S. manufactured *Globalstar* satellites on a single LM-2E rocket and 22 Iridium satellites on 11 LM-2C vehicles. CGWIC and Hughes Space and Communications Company have struck a long-term deal under which China will launch 10 satellites through the end of 2006.[10] Assuming no serious rash of failures, the number of planned Chinese commercial launches total 27 in the next 7 years, approximately $1.35 billion in revenue.[11]

Other Space Transportation Programs.

COSTIND and CASC will need this cash flow to augment budgets for a wide variety of planned space programs. The decision to proceed in R&D into manned space platforms and reusable space vehicles was made in formulating the SSTC's Mid- to Long-Term S&T Development Program.[12]

- Space Capsule (*Feichuan*). China has been conducting research into manned space flight since 1970s. Under Project 921, China is planning to launch two astronauts into space in 1999 for 5 days to celebrate the 50th anniversary of the founding of the communist state. The initial test of the KM-6 space capsule, roughly comparable to the two-man U.S. *Gemini* spacecraft, will be launched in 1998, probably from Jiuquan Space Launch Center in Gansu province. The launch vehicle will most likely be a variant of the LM-2E. Under a 1995 contract with CAST, Russia's Yuri Gagarin Center near Moscow is providing training for 70-80 Chinese astronauts, engineers, and managers in 1997-98. Russia's Krunichev Space Center and Energia Company and the Ukrainian space agency are assisting in the development of the capsule and booster capable of lifting 20 tons.[13] China's domestic astronaut training has been underway at COSTIND's Beijing Institute of Space Medical Engineering (507th Research Institute) since 1968. A new astronaut training facility, the Beijing Space Technology Experiment Center, was opened in 1995 in northwestern Beijing.

All astronauts in training are experienced pilots with at least 1000 hours of flying time.[14]

- Space Shuttle (*Hangtian Feiji*). Since 1989, China has embarked upon a serious effort to deploy a space shuttle. With COSTIND and CASC in the lead, Chinese aerospace engineers are absorbing technologies and lessons learned from the Soviet and U.S. experiences with space shuttles. In fact, Russia is granting significant assistance to China's space shuttle.[15] Space shuttle designs are somewhat alarming. According to one U.S. analysis, a 1991 design was meant to optimize transfer between coplanar orbits, essential for military related space activities to include ASAT operations. Other Chinese studies confirm interest in coplanar transfers.[16] Some observers note that the first experimental launch could occur as early as 2005. The project is valued at RMB 11 billion (approximately $1.35 billion in U.S. dollars), weighs 22 tons, carries a payload of up to 3.5 tons, and is operated by a crew of three. The shuttle will have a service life of 30 missions, with a typical mission length being 3-5 days.[17]

- Aerospace Plane (*Kongtian Feiji*). Taking space shuttle concepts a step further, COSTIND is directing an effort to master technologies associated with a hypersonic single stage-to-orbit (SSTO, or *danji rugui*) aerospace plane. COSTIND's Beijing Institute of Systems Engineering (BISE), with technical assistance from the launch vehicle/ballistic missile and cruise missile industries (CASC's First and Third Academies), is responsible for the aerospace plane's systems design, which will incorporate scramjet engine (*chaoran chongya fadongji*) technology.[18] First Academy President Li Jianzhong indicated one design concept weighs 2,000 tons and will have manned and unmanned versions.[19] Possibly in preparation for testing, COSTIND's China

Aerodynamic Research and Development Center has upgraded its hypersonic wind tunnel complex in Mianyang, outside of Chengdu, Sichuan province. The wind tunnel will be able to test various systems traveling at Mach 5-10.[20] The First Academy (11th Research Institute), Shanghai Academy of Space Technology (Institute of Power Machinery), and the Beijing University of Aeronautics and Astronautics have also proposed other aerospace launch vehicle concepts.[21]

- Tethered Satellites (*Xisheng weixing*). China has initiated R&D into space tether technology for generating power and in the repositioning of satellites. A tether system reels out what is in effect a small satellite which can generate a large amount of electricity. Tethers have generally been used on U.S. space shuttle systems.[22]

- Space Station. China is laying the groundwork for a space station around the year 2020.[23]

- Moon Missions, Mars Probes, and Deep Space Exploration. China's space community is also planning to launch a scientific satellite to orbit the moon around the year 2000. This scientific satellite will serve as the initial step in a long-range program in lunar exploration. Specific approaches to exploration of the moon includes an unmanned lunar surveyer as well as landing a man on the moon. China's space community is evaluating feasibility of a Mars probe, and also intends to field a initial pair of satellites for scientific exploration of deep space.[24]

Satellite Development.

Since the 1970s China has fielded seven COMSATs, two METSATs, 17 retrievable photoreconnaissance satellites, and 12 other scientific systems. One of China's objectives is to eliminate reliance on foreign communications, weather, navigation, and photo-reconnaissance satellites. For example, as of June 1996, China depended on foreign satellites for up to 80 percent of its satellite communications. Besides reducing vulnerabilities related to foreign dependence, COSTIND, CASC, and MPT view marketing of communications and other satellites as an extraordinary source of revenue. Besides renting transponder space, CAST is formulating plans to sell satellites by 2000.[25]

The general direction of China's satellite development is to reduce the number of components and reduce the weight of their satellite systems. Power supplies make up to 40 percent of the weight, satellite structure around 30 percent, and attitude control system around 20 percent.[26] CAST is striving to develop larger antennas, multiple beams, onboard processing, and more capable power sources. For better satellite control, the Fifth Academy (CAST) has integrated GPS onto their satellite platforms. After initiating preliminary research in 1991 and model R&D in 1994, the 503rd Research Institute tested its first GPS system on board China's 17th remote sensing satellite launched in October 1996.[27]

To expand the lifespan of their space systems, the PLA and the space and nuclear industry have advocated using a nuclear generator to power satellites and other platforms in space. Some preliminary research had been done already between 1970-78 when the Institute of Atomic Energy and COSTIND jointly carried out research on space reactors. In May 1994, 50 military and civilian experts from various entities participated in a space nuclear power technology seminar. At the conclusion, participants petitioned the State Council to allocate increased funding for nuclear powered satellites in order to place a trial nuclear powered satellite into orbit within 10 years. The group strongly argued for cooperating with foreign space industries.[28] Since 1988, China has been carrying out research into arcjet satellite engines which provide an efficient manner of utilizing propellant.[29]

In their satellite development, China is working toward launching their second generation communication satellites which will be a vital component to their national information infrastructure. With General Staff Department (GSD) and COSTIND guidance, the space industry is working toward fielding a full constellation of reconnaissance systems, including high resolution EO, radar, and infrared systems, which, in the outyears, could give them a near real time imaging and missile early warning capability. The space industry is also working to field their second generation weather satellite and first generation navigational satellites.

CAST has had significant problems, however, in its satellite development. The satellite R&D cycle is usually 7-8 years long, and operating lives of less than 4 years are too short; resolution of remote sensing platforms is low; and capacity of communications satellites are low. Only about one-third of the space industry is devoted to advanced research, design selection studies; and engineering management.

China's communications and remote sensing satellites are discussed in detail in the main body of this report. Besides communications, scientific, and remote sensing platforms, CASC is also working on fielding a sophisticated weather satellite. The new generation weather satellite, the *Fengyun-2* (FY-2), exploded at the Xichang Space Launch Center. After completion and launch of a replacement, China expects to complete the FY-3 follow-on. FY-2 will be positioned in a geosynchronous orbit over Singapore (105 degrees east).[30] Weather satellites are essential for the conduct of a variety of military operations.

CASC is also carrying out R&D on navigation satellites. Chinese exploitation of GLONASS and NAVSTAR GPS is well-documented. China, however, loathe to be dependent on any foreign system, is busy developing its own navigation satellite constellations. The initial design, the Twinstar (*shuangxing*) Rapid Positioning System, is an RDSS design with two satellites in geosynchronous orbit, due for launch around the year 2000. The follow-on system under design consists of four satellites also in a geosynchronous orbit.[31]

Unlike GLONASS and NAVSTAR GPS, these geosynchronous systems will only provide regional navigation services, and will probably not provide enough precision for use on missile systems. CASC, however, has been instructed to proceed with design of a more complex global navigation system. CAST's systems engineering design institute has forwarded two conceptual designs which would provide global coverage. One design employs a constellation of five satellites in five orbital planes at an inclination of 43.7 degrees. An alternative calls for seven satellites in seven orbital planes at a 61.8 degree inclination.[32]

CAST is also engaged in preliminary R&D into even more advanced satellite systems. For example, CAST has an active R&D program for a data relay satellite, crucial for a global near real-time imaging capability. They are also conducting conceptual studies on a space-based satellite tracking system which would serve as a potentially important component of any ASAT system.[33]

Satellite Miniaturization.

China has embarked on a program to develop a family of small satellites as a measure to reduce costs, increase revisit rates, and increase survivability. General classifications of these satellites include small satellites (100-500 kilograms); microsatellites (10-100 kilograms); and nanosatellites (less than 10 kilograms). Critical steps in developing small satellites include smaller, more efficient power sources, smaller on-board computers and attitude control systems, and reduced structure size.[34]

The Chinese clearly recognize the military implications of small satellites. Chinese defense officials advocate small satellite development in order to reduce vulnerability of fixed launch sites. Chinese engineers are examining the utility of using mobile, solid-fueled launch vehicles, such as a modified DF-21, or future variants of the DF-31 and DF-41. They are also interested in satellite launches from transport aircraft.[35] Reduced size and complexity allows for faster R&D and manufacturing time, and production in significant numbers.

In a contingency situation, tactical communications and imagery satellites can be launched on demand. Mobile launch platforms allow for increased survivability. As seen by China's Globalstar contract, multiple small satellites can be launched on a single launch vehicle. Furthermore, ASAT attacks on small satellite constellations will encounter greater targeting difficulties and be costly. Destruction of one satellite will have minimal effect on the overall functioning of the constellation.[36]

Preliminary research has been conducted on small satellite photo-reconnaissance and electronic reconnaissance constellations. CAST is also examining future generation small satellite communications, satellite constellations which are put into low earth, sun synchronous orbit, transmit via a digital storage and forward system, incorporate spread spectrum, and other secure methodologies.[37]

One of the first family of small satellites is the *Shijian* (SJ) series. The SJ-4 was a small satellite experiment which included a package of various scientific experiments. With 863 Program funding and COSTIND oversight, the SJ-4 weighed 400 kilograms and, after launch by a LM-3A, was placed in a 200-36000 kilometer geosynchronous transfer orbit. CAST provided the bus while CAS's Space Science and Application Research Center provided the payload. The follow-on SJ-5 system is expected to be the common bus for China's first generation of small satellite constellations. SJ-4 was the fourth in a series of scientific platforms. SJ-1 was launched in March 1971, while SJ-2, SJ-2A, and SJ-2B were three space physics satellites launched on a single launch vehicle in September 1981. SJ-3 was designed but aborted.[38]

Foreign Assistance.

Beijing clearly understands China cannot develop a viable space capability without significant foreign technical assistance. After establishing a relationship with Germany and France in the 1980s, CASC has expanded its contacts with a number of countries, including the former Soviet Union, Brazil, and the United States.

The greatest source of foreign assistance for its space program is from countries of the former Soviet Union. Space cooperation between China and Russia has been formalized into a series of agreements between CASC and the Russian space agencies. Representatives from Chinese and Soviet space industries signed an initial agreement in Moscow in May 1990 on 10 cooperative projects, the first of which addressed joint efforts to develop a GLONASS/GPS compatible receiver. A formal contract was signed 2 years later. The relationship was solidified when, on December 18, 1992, CASC and the Russian Space Agency signed an official protocol for the sharing of space technology. This agreement was formalized during President Yeltsin's visit to Beijing where he also signed a no-first-use pledge with the Chinese.

A follow-on agreement, signed by CASC Vice-director Wang Liheng and the head of the Russian Space Agency, outlined at least ten areas of space cooperation including exchanges in satellite navigation, space surveillance, propulsion, satellite communications, joint design efforts, materials, intelligence sharing, scientific personnel exchanges, and space systems testing.[40] For program management, the two sides agreed to annual meetings to review the various cooperative programs. In addition, China, Russia, and Ukraine's space community have held annual exchange conferences attended by as many as 250 engineers. The fourth annual trilateral conference was held in September 1996 in Kiev, Ukraine.[41] Space cooperation agreements have also been concluded with Ukraine (March 1994), Belarus (June 1994), and Kazakhstan (May 1998). Areas of cooperation with Ukraine include remote sensing, satellite communications, and aerospace material research and development.[42]

Much of the Sino-Russian and Sino-Ukrainian cooperation is centered on China's manned space program and supporting launch vehicles. Russia and Ukraine have provided assistance to China in developing cryogenic upper stages for launch vehicles. Russia sold three RD-120 cryogenic upper stage engines, the same as that used on the Zenit space launch vehicle, to China.[43] A formal agreement for Russian assistance in the manned space program was slated to be signed during Li Peng's visit to Moscow in May 1995 or during Yeltsin's late 1995 visit to Beijing. China is allegedly paying "billions of dollars" for Russian assistance in its manned space program, although a small proportion can be paid in bartered goods. Items of special interest include emergency rescue, thermal control, and space docking.[44]

China has not limited its international cooperation to the former Soviet Union. Building upon a foundation established in the 1980s, recent contacts between French and Chinese space communities were initiated during June 1994 space industry negotiations between France's National Center for Space Studies (CNES) and COSTIND and CASC. Specific areas of discussion included small launch vehicle technology and R&D of small satellites. Two follow-up meetings were held in September 1994 and November 1995. Since 1994, at least 10 Chinese space delegations have visited French space industry facilities. Besides small launch vehicle and satellite technology, cooperation focuses on GPS/GLONASS exploitation, satellite attitude control systems, communication satellites, and meteorological satellite technology.[45] China will work with France in developing the Proteus small satellite bus.[46]

The space relationship between China and Germany started in the early 1980s, culminating in a contract between the CASC and DASA in 1987 on the DFH-3. A follow-up agreement between CASC and Deutsche Aerospace AG was signed in November 1993 which directed the establishment of a joint venture between EuraSpace and CASC called Sinosat. Sinosat's first satellite is Nahuel, a communications satellite scheduled for launch in 1997.[47] Other areas of cooperation include SATCOM transponder technology, solar panels, and orbital control systems.[48] China is also working with Germany on a two ton solar telescope which will orbit around the moon about 2002.[49]

Space Warfare.

Chinese strategists have grasped the concept of space dominance. Writings and speeches by the PLA leadership acknowledge space as an essential dimension of regional warfare. Space systems, under the responsibility of CASC's CAST, are emphasized in China's 863 program and the 5- and 15-year plans. These systems include a new generation of COMSATs, data relay satellites, remote sensing platforms, a navigation satellite constellation, weather satellites, space shuttle, and space station. Chinese space advocates place special emphasis on small satellite technology.

Chinese writings indicate a significant interest in anti-satellite warfare, an issue usually shrouded in extensive secrecy.[50] When discussing ASAT principles, internal writings omit sections which address China's domestic programs. However, a number of systems now under development could be used in an ASAT role. First, China has an extensive satellite tracking network which may be upgraded to provide precision tracking for space intercepts. Sites are located throughout the country, and COSTIND has approached a number of foreign countries, including Chile and Brazil, to jointly build and use space tracking facilities. Besides tracking and controlling satellites, these sites could receive downlinked imagery for transmittal to Beijing.

With some modification, China's solid fueled missile systems, the DF-21 or DF-31, could be used in a direct ascent ASAT role. One Chinese article referred to the potential use of an older launch vehicle, the *Long March-1*, in an ASAT mode. Warheads would most likely rely on kinetic or RF energy. The successful development of a ballistic missile defense system will give the Chinese an inherent ASAT capability as well.

Other ASAT concepts which have appeared in Chinese writings include a satellite which can spread steel balls along an enemy satellite's path or spread dust on satellite reflectors. Finally, COSTIND has acknowledged research into the use of enhanced radiation weapons, which could be most useful in a nuclear pumped laser system. China probably has the technological capability to jam satellite uplinks or downlinks with some effectiveness, and could use ground-based lasers for opto-electronic countermeasures.

Chinese strategists are thinking about satellite defense as well. Satellite hardening measures, such as thermal shielding or armor, are likely being examined by COSTIND and CAST. Deception or hiding measures could be employed as well. The use of small satellites, as CASC analysts have acknowledged, will reduce satellite vulnerability. The use of gallium arsenide (GAAS) integrated circuits will provide additional resistance against space EMP weapons. China's most recent photoreconnaissance satellite, the FSW-2, displayed some in-orbit maneuvering ability. As a side note, one cannot dismiss the possibility that Russian institutes have provided some results from previous Soviet ASAT tests.[51]

Finally, China's increasing reliance on internationally shared space platforms could complicate attempts to deny access to or destroy space systems used by the Chinese. China and Brazil have a joint project to deploy two remote sensing satellites. Beijing utilizes transponder space on the Hong Kong controlled Asia-Pacific Telecommunications satellite. Besides complicating ASAT efforts, the joint development and use of satellites reduces the cost of the wide range of satellite development programs.

Who should oversee China's fledgling space force is a subject of internal debate.[52] The most likely candidate for running a space organization is COSTIND's China Launch and Tracking Control (CLTC). CLTC oversees China's three main launch centers and its satellite tracking and control network. There have been discussions within the PLA leadership on splitting off CLTC into a independent space command, or under control of the PLAAF, but there has been no concrete action to date.[53]

In summary, China has a wide variety of programs which indicate it intends to become a major player in space in the 21st century. China's emphasis is on dual use systems which could provide significant value added benefits to future PLA operations. PLA strategists understand that without a presence in space, national security cannot be assured.

Foreign assistance is crucial in their efforts. Launch vehicles, besides providing an important source of revenue, serve as the foundation of their future in space. A broad family of satellite systems will provide critical C^4I services for their emerging reconnaissance/strike complex. For greater space survivability, greater production capacity, and lower costs, China is emphasizing small satellite development. Finally, extensive R&D is being carried out on a counterspace capability in order to potentially deny future adversaries access to their own space systems.

ENDNOTES - APPENDIX III

1. Wang Chunyuan, *China's Space Industry and Its Strategy of International Cooperation*, Stanford University Center for International Security and Arms Control, July 1996, p. 16.

2. Author's 1994 visit to 211 Factory, Weinan Satellite Control Site, and Xian Satellite Control Center; Zhang Rufan, "Development and Prospects for China's Space Technology," *Zhongguo Hangtian*, June 1994, pp. 6-10, in *JPRS-CST*-94-019.

3. Chang Yuyan, "New Launch Forms for Spacecraft," paper presented at the China Astronautics Society Conference on Launch Engineering and Ground Equipment, July 1994, 5 pp.

4. Zhang Xinzhai, "Achievements and Future Development of China's Space Technology," *Zhongguo Hangtian*, June 96, in *Foreign Broadcast Information Service-CST* (hereafter *FBIS-CST*)-96-015.

5. *Ibid*. Also see Chen Lan, "Chinese Manned Space Program," in *Dragon in Space*, an internet site managed out of Singapore, April 10, 1998.

6. Craig Covalt, "Chinese Manned Flight Set for 1999 Liftoff," *Aviation Week and Space Technology Review*, October 21, 1996, pp. 22-23. "Ambitious Space Program Gears Up for Competitive Edge," in *Xinhua*, October 31, 1996, in *FBIS-CHI*-96-215; and "Scientist Urges Developing Large Launch Vehicles," *Xinhua*, March 22, 1998, in *FBIS-CHI*-98-081.

7. Covalt, pp. 22-23.

8. Chen Chunmei, "Long March Booked For Two More Satellites," *China Daily*, in *FBIS-CHI*-96-146, July 29, 1996, p. 1.

9. Based on author's discussions with various COSTIND and CASC space officials, 1992-95. CASC's annual expenditure dual use space R&D and production amounts to RMB 400 million, at an exchange rate of 8.5, which equals U.S. $47 million. Total state space budget is around $100 million (0.035 percent of overall Gross National Product). On average, a launch vehicle or missile costs between 3.5 and 10 million U.S. dollars, and a satellite costs between 5 and 12 million. See "Luan Enjie Fuzongjingli Tan Hangtian Zhiliang," *Zhongguo Hangtian Bao*, March 21, 1994, p.1; and author's unpublished report, "The Role of China's Space and Missile Industry in PLA Strategic Modernization," May 1995.

10. "China to Launch 10 Satellites For Hughes Through 2006," *Xinhua* News Release, June 21, 1997.

11. Total based on a conservative average of $5,000 U.S. dollars per launch. See Wang Chunyuan, p. 21.

12. Si Liang, "Only the East Wind is Lacking," *Zhongguo Tongxun She*, August 11, 1994, in *FBIS-CHI*-94-158.

13. Craig Covalt, "Chinese Manned Flight Set for 1999 Liftoff," *Aviation Week and Space Technology Review*, October 21, 1996, pp. 22-23; and "Chinese Cosmonauts To Train at Gagarin Space Center," *Interfax*, December 27, 1996, in *Foreign Broadcast Information Service-Soviet Union* (hereafter *FBIS-SOV*)-96-250. For indications of Jiuquan serving as the base for China's manned launch program, see Xu Kejun, "Safety Support of Space Launching For Astronauts," *Daodan Shiyan Jishu*, 1997 (1), in *CAMA*, 1997, Vol. 4, No. 5, pp. 1-9.

14. *China Today: Defense Science and Technology*, Beijing: National Defense Industry Press, 1993, pp. 386-390; and "Astronauts Trained for Chinese Manned Spacecraft," *Zhongguo Tongxun She*, March 29, 1995, in *FBIS-CHI*-95-062.

15. "China Dedicates Itself to the Research of Space Shuttle, to Send Men Into Space in the Next Few Years," *Wen Wei Po*, September 24, 1994, p. A1, in *FBIS-CHI*-94-204.

16. For a technical analysis of one of China's space shuttle concepts, see Philip A. Smith, "The Chinese Space Shuttle: A Great Leap Forward," unpublished Naval War College paper, June 1996; and Wang Xiaojun, "Study on Minimum-fuel Low Thrust Multiple Burn Transfer Between Coplanar Circular Orbits," *Yuhang Xuebao*, 1996, Vol. 17, No. 3, in *CAMA*, 1996, Vol. 3, No. 6.

17. Viacheslav A. Frolov, "China's Armed Forces Prepare for High Tech Warfare," *Defense and Foreign Affairs: Strategic Policy*, January 1998, pp. 12-13. The 2005 timeframe appears overly ambitious.

18. See Liu Xingzhou, "*Chongya Fadongji de Fazhan He Yingyong*" *(Development and Applications of Pulsed Engines), Zhongguo Hangtian*, Vol. 3, 1993, pp. 34-37; Huang Zhicheng, "Real Gas Effects for the Aerospace Plane," *Aerodynamic Experiment and Measurement Control,*" Vol. 8, No. 2, 1994, pp. 1-9; also see Huang Zhicheng, "Aerodynamic Design of Nozzles for the Aerospace Plane, "*Aerodynamic Experiment and Measurement Control*," Vol. 7, No. 4, 1994, pp. 1-7; Qiu Qianghua and Huang Zhicheng, "Aerodynamic Design Of Single Stage to Orbit Winged Rocket," *Liuti Lixue Shiyan Yu Celiang*, 1997, 11 (2), pp. 19-25, in *CAMA*, 1997, Vol. 4, No.6; Wang Yali, "Hybrid Propulsion System for a Single-Stage-To-Orbit Engine," *Huojin Tuijin*, 1997, Vol. 3, pp. 39-49, in *CAMA*, Vol. 4, No. 6; and Yue Jialing, "Preliminary Design of a Hydrocarbon-Fueled Dual Mode Scramjet Engine for a Hypersonic Vehicle," *Liuti Lixue Shiyan yu Celiang*, 1997, Vol. 11, No. 2, pp. 1-13, in *CAMA*, Vol. 4, No. 6. China's interest in an aerospace plane is not unusual. Almost all the major powers, including the United States, Europe, Japan, and India, have active hypersonic vehicle programs.

19. Presentation delivered at 47th International Astronautical Federation, March 1996, Beijing, China.

20. Liu Xiaojun and Liu Jun, "Nation's Largest Hypersonic Wind Tunnel Gets Facelift," *Keji Ribao*, July 24, 1996, p. 2, in *FBIS-CST*-96-017.

21. Qi Feng, "Preliminary Analysis on Single Stage to Orbit Launch Vehicle," *Tuijin Jishu (Propulsion Technology)*, 1996 17 (4), in *CAMA*, Vol. 3, No. 6; Zhang Xuehe, "Investigation of Propulsion Systems For SSTO Rocket Engines," *Hangtian Qingbao Yanjiu*, HQ-96023, in *CAMA*, 1997, Vol. 4, No. 2; and Nan Ying, "Optimal Ascent Trajectories for Horizontal Take Off and Landing Two-Stage-to-Orbit Aerospace Plane," paper presented at the 7th Meeting of the National Space and Launch Vehicle Control Technology Society, September 1996, in *CAMA*, 1997, Vol. 4, No. 3; and Nan Ying, "Optimum Trajectory and Propulsion Chacteristics of An Aerospace Plane," *Feixing Lixue*, 1997, Vol. 15, No. 2, pp. 27-32, in *CAMA*, Vol. 4, No. 6.

22. Zhu Renzhang, "On the Use of Tethers for Re-entry of Space Capsules," *Yuhang Xuebao (Journal of Astronautics)*, 1994, Vol. 15, No. 4, pp. 24-30, in *CAMA*, 1995, Vol. 2, No.1; Lin Huabao, "Development and Application of Space Tether Technology," *Kongjian Jishu Qingbao Yanjiu*, July 1994, pp. 314-341, in *CAMA*, Vol. 1, No. 6; Gou Xingyu,

"Research on Tethered Satellite System," *Hangtianqi Gongcheng*, 1995, Vol. 4, No. 4, pp. 34-42, in *CAMA*, Vol. 3, No. 3; and Cui Naigang, "Calculation of Orbital Elements For Released Tethered Satellite," *Journal of Harbin Institute of Technology*, April 1997, Vol. 2, pp. 126-130, in *CAMA*, Vol. 4, No. 6; and Yu Shaohua, "Periodic Motion of a Tethered Satellite System," *Yuhang Xuebao*, 1997, Vol. 18, No. 3, pp. 51-58, in *CAMA*, Vol. 4, No. 6.

23. "China Preparing for Manned Space Program," *Neimenggu Ribao*, July 28, 1995, p. 2, in *FBIS-CST*-95-015.

24. "Beijing Preparing for Moon Mission," *Xinhua*, January 13, 1995; Chen Jianxiang, "Approaches to Earth-to-Moon Trajectories," *Kongjian Kexue Xuebao*, 1997, 17 (1), pp. 90-96, in *CAMA*, 1997, Vol. 4, No. 2; Chen Ji and Li Xiguang, "PRC Scientist on Lunar, Mars Probe Feasibility Studies," *Xinhua*, March 20, 1998, in *FBIS-CHI*-98-079; Chen Ji and Li Xiguang, "China Considers Venturing Into Space Exploration, *Xinhua*, March 20, 1998, in *FBIS-CHI*-98-082; and Zeng Guoqiang, "Study on Soft Landing Orbit of Lunar Detector," *Guofang Keji Daxue Xuebao*, 1996, 18 (4), pp. 40-43, in *CAMA*, 1997, Vol. 4, No. 2. As expected, COSTIND appears to have the lead in R&D on China's lunar exploration program.

25. Ding Henggao, "Speeding Up Development of Space Industry To Meet New Challenges," *Zhongguo Hangtian*, June 1996, in *FBIS-CST*-96-015; Zhang Rufan, "Prospects and Development of China's Space Technology"; and "Official On Plans to Export Satellites by Year 2000," *Zhongguo Xinwen She*, November 16, 1996, in *FBIS-CHI*-96-223.

26. Zhang Xinzhai, "Achievements and Developments of China's Space Technology," *Zhongguo Hangtian*, June 1996, in *FBIS-CST*-96-015.

27. Liu Zhen, *"Gaodongtai GPS Jieshouji Zai Weixingshang Tazai Chenggong"*(GPS Receiver Carried on Satellite Successful), *Zhongguo Hangtian Bao*, November 16, 1996, p. 1.

28. Russia is the obviously best candidate for foreign cooperation. See Yang Qifa, "Research and Applications of Space Nuclear Power," *Hangtianqi Gongcheng (Spacecraft Engineering)*, Vol. 4, No. 4, 1995; also see "Experts Urge Research on Space Nuclear Power Supply," *Zhongguo Tongxun She*, June 17, 1994, in *FBIS-CHI*-94-138 and Li Huailin, "Chinese Research on Space Nuclear Reactors Highlighted," *Keji Ribao*, August 29, 1996, p. 7, in *FBIS-CST*-96-017.

29. Wu Hanji, Feng Xuezhang, and Liu Wenxi, "Experimental Investigation of Arcjet Engines," *Zhongguo Kongjian*, August 1997, pp. 41-48, in *FBIS-CHI*-98-111. State funding began in 1992.

30. Zhang Xinzhai, "Achievements, Future Development of China's Space Technology," *Zhongguo Hangtian*, June 1996, in *FBIS-CST*-96-015.

31. Zhong Xiaojun, "Synchronous Satellite Navigation on the Large Regional Earth Surface and in the Near-Earth Space," *Kongjian Jishu Qingbao Yanjiu*, July 1994, pp. 243-253, in *CAMA*, Vol. 1, No. 6. Zhong is from the Lanzhou Institute of Physics.

32. Yu Mingsheng and Dai Chao, "Design of Satellite Constellations," paper presented at the China Space Society, Ninth Meeting of Space Surveying, November 1996, in *CAMA*, 1997, Vol. 4, No. 3. The satellites are inserted into orbit at 72 degrees and 51.4 degrees.

33. Cheng Yuejin, "Information Transmission System of Data Relay Satellites," *Kongjian Jishu Qingbao Yanjiu*, July 1994, pp. 185-193, in *CAMA*, Vol. 1, No. 6. Cheng is from the Xian Institute of Radio Technology. Also see Tan Liying, "Selection of Wavelength Region for Optical Intersatellite Communication," *Haerbin Gongye Daxue Xuebao*, 1994, Vol. 26, No. 3, pp. 24-27, in *CAMA*, Vol. 1, No. 6; Chen Daoming, "Frequency and Orbit of Data Relay Satellites," in *Zhongguo Kongjian Kexue Jishu*, 1996, Vol. 16, No. 1, pp. 26-31, in *CAMA*, Vol. 3, No. 3.

34. Wang Jingquan, "*Xiaoweixing Jiang Yinqi Kongjian Jishu Fazhan de Yichang Geming*," (Revolution in Small Satellite in Space Technology Development), *Zhongguo Hangtian*, September 1994, pp. 31-33; Zhang Guofu, "*Xiaoxing Weixing Qishilu*" (Accomplishments in Small Satellites), *Zhongguo Hangtian*, June 1992, pp. 23-27.

35. These concepts has been closely examined and strongly advocated by the space and missile industry. See Zhang Dexiong, "*Guowai Xiaoxing Weixing de Guti Huojian Tuijin Xitong*" (Solid Rocket Propulsion Systems for Foreign Small Satellites), in *Hangtian Qingbao Yanjiu*, HQ-93011, pp. 139-155; Wang Zheng, "Screening Studies and Technology for All-Solid Space Launch Vehicles," *Guti Huojian Fadongji Sheji Yu Yanjiu* (*Solid Rocket Engine Design and Research*), April 1996, pp. 63-73, in *CAMA*, 1996, Vol. 3, No. 6; Zhang Song, "Design and Optimization of Solid Launch Vehicle Trajectory," *Guti Huojian Jishu*, 1997, 20 (1), pp. 1-5; and Zhang Dexiong, "China's Development Concept for Small Solid Launch Vehicles," CASC Fourth Academy

Information Research Reports, the Fourth Edition, October 1995, pp. 1-11, in *CAMA*, Vol. 5, No. 2.

36. Mao Genwang and Wang Liang, *"Weixing de Junshi Yingyong Tedian, Fazhan Xianshi yu Yingyong Qianjing"* (Key Applications of Microsatellites and Prospects for Development), *Zhongguo Hangtian*, May 1992, pp. 33-53.

37. Wang Mingyuan, "Digital Storage and Forward Communication Small Satellite," *Kongjian Dianzi Jishu (Space Electronics Technology)*, February 1996, pp. 23-36, in *CAMA*, 1996, Vol. 3, No. 6. The ELINT and remote sensing constellations are discussed in the main report.

38. "Ambitious Space Program Gears Up for Competitive Edge," and Zhu Yilin, "China's Shijian-4 Satellite," unpublished paper.

39. Wang Chunyuan, *China's Space Industry and Its Strategy of International Cooperation*, Stanford University Center for International Security and Arms Control, July 1996, p. 4; Marat Abulkhatin, "Official on Prospects for Space Cooperation," *Itar-Tass*, October 10, 1996, in *FBIS-SOV-96-198*.

40. *"Wang Liheng Fujuzhang Lutuan Fangwen E'Wu Liangguo"* (CASC Deputy Director Wang Liheng Leads Delegation to Russia and Ukraine), *Zhongguo Hangtian Bao*, April 11, 1994, p. 1.

41. Chen Chang, "Ukrainian Deputy Prime Minister Says There Are Broad Prospects for Ukrainian-Chinese Cooperation," *Xinhua*, September 17, 1996, in *FBIS-CHI-96-182*.

42. *"Zhong, E, Wu, Disanjie Yuhang Keji Dahui Zai Xigongda Zhaokai"* (China, Russia, Ukraine, Open the Third Space Science Meeting at Northwest Polytechnical University), *Zhongguo Hangtian Bao*, October 3, 1994, p. 1. The 1994 meeting was held from September 17-20. Sun Jiadong chaired the Chinese team for the Sino-Kazak.

43. "Russia Sells Launch Vehicle Technology," *Military Space*, August 21, 1995.

44. Andrey Kirillov, "China to Buy Russian Equipment for Manned Space Missions," *Itar-Tass*, March 14, 1995, in *FBIS-SOV-95-049*.

45. Christian Lardier, "Chinese Space Industry Ambitions," *Air & Cosmos/Aviation International*, October 25, 1996, in *FBIS-CST-96-020*; and author's unpublished report, "The Role of China's Space and Missile Industry in Strategic Modernization," May 1995.

46. "Paris, Beijing Discuss Strategic Space Partnership," *Les Echos*, January 29, 1997, p. 11, in *FBIS-CHI*-97-020.

47. Liu Jiyuan, "Strengthening Space Cooperation, Looking Forward to the 21st Century," *Zhongguo Hangtian*, June 1996, pp. 7-8, in *FBIS-CST*-96-015.

48. *"Zhongguo Hangtian Kaizhan Quanfangwei Guoji Hezuo"* (China Space Opens Comprehensive International Cooperation Meeting), *Zhongguo Hangtian Bao*, October 3, 1994, p. 1.

49. "Ambitious Space Program Gears Up for Competitive Edge."

50. ASAT development is discussed in detail in the main body of this report.

51. For a discussion of the increased need for space system defense, see Ping Fan and Li Qi, "A Theoretical Discussion of Several Matters Involved in the Development of Military Space Forces," *Zhongguo Junshi Kexue (China Military Science)*, May 20, 1997, pp. 127-131. Ping and Li are from COSTIND's Command Technology Academy.

52. For an overview of the debate, see Ping Fan and Li Qi article on development of space forces.

53. Frolov, p. 13.

APPENDIX IV

CHINA'S DIRECTED ENERGY WEAPONS

China is examining and testing the critical engineering technologies needed to develop a host of directed energy weapons (DEW). The world, including China, has long been fascinated with the notion of projecting energy at the speed of light in the pursuit of warfare. American and former Soviet Union DEW programs are well-known. Very little, however, is publicly known about research and development (R&D) of DEW in China. However, the number of papers on directed energy weapon-related technology is in itself an indication of the amount of work done in this field. DEW are part of a larger class of weapons known to the Chinese as "new concept weapons" (*xin gainian wuqi*), which include high power lasers, high power microwaves, railguns, coil guns, particle beam weapons, etc.

Organization of China's DEW Program.

The two most important organizations involved in R&D of DEW are the China Academy of Sciences and the Commission of Science, Technology, and Industry for National Defense (COSTIND).

China Academy of Sciences (CAS). CAS is China's leading institute for basic research into the basic sciences. Institutes involved in DEW affiliated research include the following:

- Shanghai Institute of Optics and Fine Mechanics (solid state lasers; adaptive optics)

- Anhui Institute of Optics and Fine Mechanics (adaptive optics)

- Institute of Optoelectronics (adaptive optics)

- Dalian Institute of Chemistry and Physics (COIL).

China Academy of Engineering Physics (CAEP). CAEP is an outgrowth of the nuclear industry's 9th Academy (Northwest Nuclear Weapons Research and Design Academy) which has developed and tested China's nuclear warheads. The 9th Academy, subordinate to COSTIND, was previously located in Haiyan, Qinghai Province, but moved to Mianyang, Sichuan province in the late 1960s. With its headquarters in Sichuan, CAEP has offices and research entities in Shanghai and Beijing.[1] CAEP's traditional purview has been the development of nuclear weaponry, and R&D efforts into DEW systems reflect a significant diversification of its mission. Subordinate research institutes include the following:

- Southwest Institute of Applied Electronics (HPM)

- Southwest Institute of Fluid Physics (accelerators and HPM drivers)

- Southwest Institute of Nuclear Physics and Chemistry (X-ray lasers)

- Southwest Institute of Nuclear Physics

- Institute of Computer Applications

Chinese Lasers and Other Directed Energy Weapons Under Development.

One area receiving a substantial amount of emphasis is lasers. Work on lasers has been going on since the 1960s, when Mao Zedong approved the 640 Plan, an advanced weapons development project focused on missile defense. The high powered laser portion was designated as Project 640-3 and was intended to counter high altitude bombers and reconnaissance platforms, such as the SR-71 which was then under development. Primary responsibility was awarded to the China Academy of Sciences' Shanghai Institute of Optics and Fine Mechanics. Research was focused in two directions: laser nuclear fusion under the leadership of Deng Ximing, and laser weapons under the leadership of Wang Zhijiang.[2]

In 1979, under threat of scaling back high powered laser R&D, COSTIND's Sun Wanlin successfully convinced the Central Military Commission to maintain the pace and even raise the priority of laser development, and then represented the laser community in the formulation of the 863 Program.[3] Today, lasers, together with space, are key COSTIND-managed programs.[4] The Chinese recognize the role lasers will play in 21st century warfare and are laying the foundation for eventual weaponization of wide range of systems. The Chinese defense industrial complex, under management of COSTIND, are carrying out R&D on a variety of lasers. China's research into lasers began in the 1960s. An estimated 10,000 people, including approximately 3,000 engineers, in 300 organizations are involved in China's laser program. Almost 40 percent of China's laser R&D is for military purposes.[5]

Solid-State Lasers. Solid state lasers make use of artificial crystals. China is a world leader in crystal growth and is expected to further improve its capabilities through the use of parallel processing computers which allow for complex calculations for analysis of crystalline materials.[6] China developed one of its first high powered solid laser devices, the Shenguang-1 (Magic Light) in 1986. This high energy Nd:glass laser system, also known as LF-12, was developed by CAS's Shanghai Institute of Optics and Fine Mechanics (SIOFM) reportedly has an output power of one terawatt.[7] The Shenguang-II Nd:Glass laser has an output power of two terawatts, and will be a major facility for China's inertial confined fusion program.[8] China Academy of Science's Shanghai Institute of Optics and Fine Mechanics (SIOFM) has developed an Nd:YAG laser with an output power of 62.5 MW.[9] An Nd:Glass laser, the Xingguang-II, has also been developed and tested. SIOFM, together with China's University of Science and Technology, developed a tunable titanium-sapphire laser with an output power of 650MW.[10]

Free Electron Laser (FEL). Chinese defense S&T community views free electron lasers (*ziyou dianzi jiguang*) as having a number of advantages, including their adjustable wavelength and bandwidth and their potential range of 5000

197

kilometers. CAEP began holding meetings on FEL development in 1985. China's first FEL, the Shuguang-1, was activated in May 1993 and achieved a power of 140 MW in July 1994, with a theoretical maximum output of 10GW. The Shuguang-1 was developed by CAEP's Southwest Institute of Fluid Physics (SWIFP).[11] The size of FEL systems currently limit deployment options. However, MEI's Southwest Institute of Electronic Engineering is working to reduce the size of FEL systems through the miniaturization of electronic devices.[12] A second FEL facility located in Beijing was developed by the China Academy of Sciences Institute of High Energy Physics.[13] Chinese FEL systems, including Raman, Compton, electromagnetic wave pumped, and Cherenkov lasers, have been modeled and simulated by an organziation with close links to CAEP, the Institute of Applied Physics and Computational Mathematics (IAPCM).[14]

X-ray Laser. X-ray lasers, in principle, can destroy electrical circuitry, possibly trigger some types of munitions, set off nuclear weapons, and make humans sick or worse. The preferred energy source is a small nuclear explosion. However, China, which claims certain aspects of its X-ray laser program to be the most advanced in the world, appears to be focusing on using high powered lasers to produce X-ray lasing.[15] China Academy of Engineering Physics (CAEP) is China's leading institute on the development of X-ray laser systems. At least one prototype X-ray driver, the Shenguang-1 device, was tested as early as 1988.[16] CAEP's Institute of Nuclear Physics and Chemistry is also working on the Xingguang-II laser to support its ICF and X-ray laser programs.[17]

Gas Lasers. China's Tianguang-1 excimer laser will probably drive CAEP's inertial confinement fusion (ICF) program.[18] CAEP's Southwest Institute of Nuclear Physics and Chemistry has developed and tested a fast breed reactor-pumped xenon laser.[19] China Institute of Atomic Energy has experimented a krypton-flouride excimer laser.[20] CAEP's Southwest Institute of Nuclear Physics and Chemistry and SIOFM have tested a nuclear pumped helium-argon-xenon laser, using their China Fast Breed

Reactor (CBFR-II). The power output reached 7.5MW with a peak power of 926MW.[21]

Chemical Oxygen-Iodine Laser (COIL). CAS' Dalian Institute of Chemistry and Physics (DICP) is responsible for China's COIL development program, which they acknowledge has the potential to be an effective DEW. COIL research in Dalian began in the early 1980s and was formally accepted as an 863 project in April 1991. Progress has been made into both pulsed and CW modes. DICP scientists claim they are among the world's leaders in COIL research.[22] In May 1995, DICP conducted one of its first tests of a COIL against a target at 140 meters, achieving a power of 10 kilowatts with an output energy of 40-50 kilojoules for 3-4 seconds.[23]

KrF Laser. China National Nuclear Corporation's Institute of Atomic Energy (IAE) has developed a krypton flouride (KrF) excimer laser. IAE began its laser development program in 1982 and on the KrF laser, named Heaven, in 1986. The laser, a potential candidate for an ICF driver, was tested in 1991 and achieved an output power of 20GW.[24]

Carbon Dioxide (CO^2) Laser. China's CO2 laser development program has been led by SIOFM which has developed a 3.2 GW system.[25] Researchers have tested the effects a CO^2 laser can have on certain composite materials, such as fiberglass epoxy.[26]

Semiconductor Laser. R&D into semiconductor lasers is a major program supported by the Ninth Five Year Plan and 2010 Long Range Plan. Semiconductor lasers have a number of advantages over conventional laser systems that make them ideal for many military applications. They are highly efficient (i.e., greater than 60 percent, as compared to 10 percent or less for conventional lasers); are scalable to higher powers; are microelectronics-based which means they can be mass produced, resulting in low cost; require low voltages, therefore, can be battery operated; are compact; and because they are solid state, they are reliable. Semiconducter lasers are also wavelength agile. A new CAS subordinated facility dedicated to the development of semiconductor lasers is being constructed in Shandong province. The Shandong Semiconductor General Plant will lead the project.[27]

Optics. China's ability to generate high powered lasers is well-established. However, a high powered laser is not enough to make a weapon system. First, optics must focus the laser energy onto a target. Generally, the larger the mirror is relative to the wavelength, the smaller the focal spot and higher the energy density. Optics must compensate for atmospheric turbulence through the use of adaptive mirrors which use numerous electronic devices to shape and achieve the optimal beam pattern. These mirrors must also be able to withstand tremendous heat generated by the laser.[28]

China has placed high priority on development of adaptive optics and deformable mirrors to support its DEW program. During an annual COSTIND science and technology (S&T) meeting, one of China's most prominent scientists and COSTIND advisor, Qian Xuesen, recognized the criticality of this technology and asserted that adaptive optics, or "magic mirrors," are one of China's most critical technologies in need of further development.[29]

Some progress in the development of adaptive optics has already been made. Concentrated efforts into adaptive optics began in 1980 when CAS formed an adaptive optics research group and laboratory within its Institute of Optoelectronic Technology.[30] CAS certified an adaptive optics system, developed by its Institute of Optoelectronics, which they believe has made China a world leader—third behind the United States and Germany—in adaptive optics.[31] CAS' SIOFM and Anhui Institute of Optics and Fine Mechanics are leading organizations responsible for development of deformable mirrors and adaptive optics.[32] Beijing's Institute of Applied Physics and Computational Mathematics (IAPCM) has conducted modeling work on atmospheric effects on ground based high power laser weapons.[33] To further boost its capabilities in optics, China is seeking assistance in laser-related optics technology from former Soviet states.[34]

High Powered Microwave (HPM) Weapons. HPM weapons are one of the world's most sought after tools of information warfare. Intended to exploit the growing reliance on dense microelectronics packed into modern military platforms, HPM weapons can degrade computer systems used in data processing systems, displays, signal processors, electronic flight controls, receivers, radars, electronic warfare equipment, satellite ground stations, and a wide range of communication systems. China—like the United States, Russia, and France—has a national program aimed at exploring the feasibility of HPM weapons. Protective measures against nuclear related EMP effects generally do not perform well against HPM weapons.[35]

China's technology base which could be applied to design of an HPM device is diverse and, in many areas, seemingly quite mature. HPM weapons are generally envisioned as part of an air defense system or, if successfully miniaturized, as a air-delivered munition, or a cruise or ballistic missile warhead. As an air defense system, HPM weapons have a number of advantages. First, there are virtually no target acquisition, pointing, or tracking requirements in any HPM employment scenario. The radiation, traveling at the speed of light, can envelop a large volume and engage multiple targets at once. It may be possible to design a system that acts as both a radar and weapon, which first tracks the target and then increases the power and engages the target.

CAEP's Institute of Applied Electronics and the Northwest Institute of Nuclear Technology in Xian are two organizations engaged in research, design, and testing of HPM devices.[36] COSTIND's National University of Defense Technology has also conducted theoretical analysis of HPM generators, as has the Southwest Institute of Electronic Equipment.[37]

Chinese military strategists, COSTIND analysts, and electronic warfare specialists believe HPM weapons will serve a useful purpose in the 21st century and strongly advocate their weaponization. HPM systems, called the "superstar" of directed energy weapons, can be used as "superjammers" which focus microwave energy onto a wide range of potential targets. HPM weapons are praised as being able to counter stealthy air targets and antiradiation missiles, and can be used against wide area targets at long distances. Other DEWs praised include plasma guns, high energy ultrasonic weapons, and subsonic wave weapons which affect the cognitive processes of a enemy pilots or soldiers. Besides HPM weapons, these include particle beam "death ray" weapons with "the power of a thunderbolt" which can carry out both hard and soft kills against ground and air targets. Space-based neutral particle beam weapons can engage missiles during all phases of flight.[38]

China's HPM program is the outgrowth of research into the effects nuclear EMP would have on its electronic equipment in a general war. Since 1964, Chinese engineers have developed a series of EMP devices to simulate ground, low altitude, and high altitude nuclear bursts. These simulators have been used to test EMP resistance of radars, C^2 centers, missile sites, underground command bunkers, and other large fixed facilities.[39]

Several power sources could drive such a weapon. For example, energy from China's FEL could be converted to a powerful electromagnetic beam.[40] Other potential generators include gyrotrons, explosively-driven magneto-cumulative generators (MCGs), magneto-hydrodynamic (MHD) generators, flux compression generators (FCG), vircators, klystons, magnetrons, or beam plasma devices, as well as backward wave oscillators. An HPM weapon could also be generated through a high explosive charge used to create a short pulse of RF energy. The explosive charge could be carried by an artillery shell, or as a warhead on a surface-to-surface or cruise missile.[41]

Numerous entities within China's defense industrial complex are conducting R&D into the most critical segment of an HPM weapon system, its power source. For example, University of Electronic Science and Technology of China (UESTC), supported by CAEP's Southwest Institute of Applied Electronics, has been conducting developmental work on a backward wave oscillator (BWO) as an HPM source since at least 1989 and acknowledges the BWO as a leading candidate for a mobile HPM system.[42] UESTC has also developed a high-powered microwave source in support of

CAEP's FEL program.[43] They have also experimented with an electron beam-plasma interaction with an HPM system which increases the operating efficiency and power.[44] UESTC, directed by Liu Shenggang, has also developed gyrotrons as an HPM source, while CAS' Institute of Electronics has been conducting R&D into broadband megawatt-class klystrons as an HPM source since 1992.[45] CAEP's SWIFP and IAPCM are carrying out R&D into a magnetic flux compression generator.[46]

Also under development at UESTC is an HPM variant known as an electromagnetic missile (*dianci daodan*), characterized by burst of RF energy which attenuates very slowly (*manshuaijian*). As a jammer, the electromagnetic missile (EMM) covers a broad frequency range from 1-40 GHZ and has a longer range than more conventional jammers. As a radar, an EMM has strong ECCM properties due to its wide bandwidth, and can effectively counter antiradiation missiles. Chinese engineers also tout its counterstealth capabilities, and ability to detect low flying targets. EMM systems can function as a most desirable means of communication due to their high capacity, long-range, ECCM properties. When boosted by a small nuclear explosion, Chinese engineers point out, an EMM can also conduct hard kills against a range of targets.[47]

One of China's first experiments in HPM weapon research was the Flash-I system which utilized a vircator as a power source. CAEP began development of the Flash-I in 1975. After completion in 1983, the Flash-I operated at approximately 1 GhZ and had a microwave power of 1GW.[48] The Northwest Institute of Nuclear Technology (NWINT) has developed the Flash-II (Shanguang) electron beam accelerator with a maximum power of 1 terawatt. After a feasibility study in 1982, the Flash-II project, designed to aid China's electromagnetic simulation and HPM weapons research, began in 1983 and was first tested in 1990.[49] NWINT has also conducted developmental work on vircators as an HPM power source and is a strong proponent for their use.[50] Finally, the 2nd Academy has also done technical assessments of directional transmissions of EMP bursts.[51]

If successful, HPM and other RF weapons could cause a military technical revolution by sharply limiting the battlespace use of advanced electronics and severely interrupting enemy communications without the use of nuclear weapons. Russia's Institute of Applied Physics and Lebedev Physics Institute are currently believed to be the world leaders in some of the basic technology involved in HPM and RF weapons.[52] In light of the large number of Russian scientists assisting China's conventional and strategic military modernization, one cannot ignore the possibility that Russian engineers are providing assistance to China's DEW program.

There is no proof or strong indication that development of HPM or RF weapons is in a more advanced stage in China than in the United States. Chinese engineers are probably working hard along the same lines as U.S. engineers on strategic and tactical applications of laser technology.

CAEP has done significant work on particle accelerator technology. Besides serving as a key component of an FEL system, accelerator technology can be used as a particle beam source, a driver for an ICF program, or to simulate a nuclear explosion. SWIFP's 3.3 MeV linear induction accelerator drives the SG-1 FEL.[53]

Potential Applications.

The PLA is placing greater emphasis on lasers and their potential military applications. The Academy of Military Science, the PLA's leading think-tank on future warfare, believes lasers will be an integral aspect of 21st century war. Strategists note a wide range of military applications of lasers, including ranging, laser radars, communications, reconnaissance, high power directed energy weapons, nuclear fusion and neutron bombs, laser gyroscopes, and laser computing. At the same time, strategists highlight the importance of countering an enemy's use of lasers.[54]

Air and Missile Defense. China's defense industrial complex is researching potential applications of a tactical laser system to counter aircraft; ballistic, cruise, and anti-radiation missiles; and a wide range of precision guidance munitions. As an air defense weapon, lasers can be highly effective against pilots and vulnerable parts of the aircraft itself, including the wing root, radar, and engine. CO^2, chemical, FEL, and X-ray laser systems have been examined as potential candidates. Research has, in large part, been budgeted under the 863 program.[55]

Satellite Tracking. For more than a decade, China has used lasers to track satellites in space. Laser range finders are located at space observatories in Wuhan, Nanjing, Beijing, Kunming, Lintong, and Shanghai. Anhui Institute of Optics and Fine Mechanics plays a large role in the development of these systems, as well as excimer lasers and the atmospheric effects on laser transmission. China's satellite laser range finders, which have an accuracy of 3-4 centimeters, assist in satellite orbit calculations and real-time tracking and adjustment in support of COSTIND's China Launch and Tracking Control (CLTC). The lasers can track satellites at both low earth and geosynchronous orbit.[56] While apparently not operationally subordinated to CLTC, the laser range finders are most likely tied into their space tracking network. The laser ranging systems can support China's domestic satellite tracking requirements and, if upgraded, any future ASAT campaign.

Anti-Satellite (ASAT). Chinese engineers have conducted at least theoretical research into the use of high powered lasers and other directed energy weapons against satellites and have closely studied U.S. and former Soviet ASAT systems, including the U.S. MIRACL, COIL, EMRLD projects, and former Soviet DEW systems at Sary Sagan.[57] Satellites as currently constructed are particularly susceptible to laser attack. Typical antisatellite concepts have a 100 second engagement time, which is the exposure time of a low earth orbit satellite to a ground station.

Laser Radar. Chinese engineers strongly advocate the development of CO^2 laser radars (lidar), especially for countering a potential adversary's cruise missiles. Supported by the 863 Program, Anhui Institute of Optics and Fine Mechanics has developed one of China's first lidars, which has a maximum detection range of 50 kilometers. Other applications include GEO satellite tracking, target tracking, fire control, and terrain following for helicopters and cruise missiles.[58] One study noted the utility of CO_2 laser radars, combined with other sensors, in countering stealthy cruise missiles and aircraft.[59]

In summary, directed energy weapons are an integral component of China's strategic modernization program. Key programs include a variety of high powered lasers and high powered microwave weapons. Chinese analysts see directed energy technology as important for China's air defense and counterspace efforts. DEW efforts also reflect a diversification of China's nuclear weapons industry.

ENDNOTES - APPENDIX IV

1. *Nuclear Weapons Databook*, Vol. V.

2. Huang He, "A Dust-Laden Secret: Review and Prospects for China's Optical Instruments and Laser Weapons," *Ch'uan-ch'iu Fang-wei Tsa-chih (Defense International)*, April 1, 1998, pp. 110-115.

3. "Achievements of the 863 Program Celebrated," *Jiefangjun Bao*, April 5, 1996, in *FBIS-CHI*-96-089.

4. Of the seven areas in the 863 Program, COSTIND views space and lasers as two of the most important. See "COSTIND Minister on Aerospace, Laser Advances," *Renmin Ribao*, April 3, 1996, in *FBIS-CHI*-96-082.

5. "Situation, Development of Laser Industry in China," *Yingyong Jiguang (Applied Laser Technology)*, June 1990, *JPRS-CST*-90-028.

6. "CAS High Performance Parallel Computing Initiative," in *FBIS-CST*-96-025.

7. "China Scores Marked Results in Laser Technology Research," *Jiefangjun Bao*, February 13, 1996, in *FBIS-CHI*-96-038. Also see "ICF, X-ray Laser Experiments With LF-12 (Shenguang) Apparatus Described," *Renmin Ribao*, May 3, 1991, *JPRS-CST*-91-019. The LF-12 was developed by SIOFM's Deng Ximing, Yu Wenyan, Fan Dianyuan, Hu Shaoyi, and Huang Zhenjiang.

8. Achievements of the 863 Program Celebrated," *Jiefangjun Bao*, April 5, 1996, in *FBIS-CHI*-96-089. As part of the ICF and X-ray laser program, CAEP opened up a lab within SIOFM.

9. "LD-Pumped Nd:YAG Microchip Laser with 62.5MW CW Output Power, *Zhongguo Jiguang (Chinese Journal of Lasers)*, May 3, 1994, in *JPRS-CST*-94-007.

10. "Ti-Sapphire Laser Attains CW Output Power of 650MW," *Zhongguo Jiguang (Chinese Journal of Lasers)*, December 1994, in *JPRS-CST*-95-004.

11. Hui Zhongxi, "Research, Design of SG-1 Free Electron Laser," *Qiang Jiguang Yu Lizishu*, August 1990, *JPRS-CST*-91-014; "COSTIND Minister on Aerospace, Laser Advances," *Renmin Ribao*, April 3, 1996, in *FBIS-CHI*-082; and "SG-1 FEL Amplifier Output Reaches 140MW," *Qiang Jiguang yu Lizishu*, December 1994, in *JPRS-CST*-94-019; Also see "SG-1 FEL, BFEL Certified; Asia's First IR-Spectrum FEL Light Generated," *Keji Ribao*, June 8, 1993, *JPRS-CST*-93-013. Performance Improvements of 3.3 MeV LIA," *Qiang Jiguang yu Lizishu*, May 1994, in *JPRS-CST*-94-014.

12. "High Performance Short-Period FEL Wiggler," *Qiang Jiguang Yu Lizishu*, May 1992, in *JPRS-CST*-92-019. SWIEE is developing a "microwiggler" as one means to reduce the FEL's overall size.

13. "SG-1 FEL, BFEL Certified; Asia's First IR-Spectrum FEL Light Generated," *Keji Ribao*, June 8, 1993, *JPRS-CST*-93-013.

14. Du Xiangwan, Dong Zhiwei, and Gao Tieda, "Recent Progress of FEL Research in China - Experiment and Simulation," paper presented at Modeling and Simulation of Laser Systems III Conference, Los Angeles, January 24-25, 1994.

15. "China Scores Marked Results in Laser Technology Research," *Jiefangjun Bao*, February 13, 1996, in *FBIS-CHI*-96-038. Also see "Bipulse Drive Generated High Gain X-Ray Laser Experiment Successful," *Qiang Jiguang Yu Lizishu*, February 1994, in *JPRS-CST*-94-008.

16. "China: 863 Program Achievements in High-Power Laser Research Highlighted," *Keji Ribao*, February 13, 1996, in *FBIS-CST*-96-013; and "COSTIND Minister on Aerospace, Laser Advances," *Renmin Ribao*, April 3, 1996, in *FBIS-CHI*-96-082. The follow-on laser for use in China's ICF program will be a neodymium glass laser, the Shenguang-2.

17. "Xingguang-II Laser Certified for ICF, X-Ray Laser Experiments," *Zhongguo Kexue Bao*, November 6, 1995, in *FBIS-CST*-96-003.

18. "China Scores Marked Results in Laser Technology Research," *Jiefangjun Bao*, February 13, 1996, in *FBIS-CHI*-96-038.

19. "Laser Cell for Reactor Pumped He-Ar-Xe Laser Experiments," *Qiang Jiguang Yu Lizishu*, August 1994, in *JPRS-CST*-94-019.

20. "Experimental Investigation of a Large-area Diode for Hundred-Joule KrF Laser," *Qiang Jiguang yu Lizishu*, May 1994, in *JPRS-CST*-94-014.

21. "Principle Tests of Reactor Pumped Laser of He-Ar-Xe System," *Qiang Jiguang Yu Lizishu*, August 1994, *JPRS-CST*-94-019.

22. "Novel High Power COIL Described," *Wuli (Physics)*, July 1992, in *JPRS-CST*-92-025; "Kilowatt-Class CW Oxygen-Iodine Chemical Laser Developed," *Zhongguo Kexue Bao (China Science News)*, September 1, 1993, *JPRS-CST*-93-017. The chemical oxygen iodine laser is the basis of the USAF's Airborne Laser which is slated for IOC around 2005.

23. Xie Ping, "Dalian Laser Achievements," *Qiang Jiguang yu Lizishu*, August 1995, Vol. 7, No. 3, back cover, in *FBIS-CST*-96-001.

24. "100J KrF Laser Pumped by Intense Electron Beam," *Qiang Jiguang Yu Lizishu*, November 1991, *JPRS-CST*-92-008. Leaders in the IAE effort include Wang Ganchang and Wang Naiyan.

25. "A GW Power Level High Power CO^2 Laser System," *Zhongguo Jiguang*, June 1990, *JPRS-CST*-90-029.

26. Wang Lijun, "Experimental Study on CO^2 Laser on Fiberglass Epoxy," *Hongwai Yu Jiguang Gongcheng*, 1996, Vol. 25, No. 2, in *CAMA*, Vol. 3, No. 4.

27. "Shandong Gets Nation's Largest Semiconductor Laser Facility," *Zhongguo Dianzi Bao*, March 18, 1997, p. 9, in *FBIS-CST*-97-010; and Han Lixin, "Shandon Builds Semiconductor Laser Production Base," *Keji Ribao*, January 16, 1997, p. 5, in *FBIS-CST*-97-005.

28. For details, see Major General Bengt Anderberg and Dr. Myron Wolbarsht, *Laser Weapons: The Dawn of a New Military Age*, New York: Plenum Press, 1992, p. 130.

29. See *Keji Ribao* (*Science and Technology Daily*), March 14, 1994.

30. "Nation's Adaptive Optics Technology in World's Front Ranks," *Zhongguo Kexue Bao*, May 21, 1991, *JPRS-CST*-91-015. The research group was led by Jiang Wenhan.

31. "Adaptive Optics Technology Is World-Class," *Keji Ribao*, October 30, 1992, in *JPRS-CST*-92-025. The adaptive optics system can also identify space targets, such as satellites.

32. "Effects of Deformable-Mirror/COAT System Finite Subaperature Size on Compensation Efficiency," *Zhongguo Jiguang (Chinese Journal of Lasers)*, February 1992, in *JPRS-CST*-92-010; and "Deomonstration of Uniform Illumination on Target by Focusing High Power Laser Beam," *Guangxue Xuebao (Acta Optica Sinica)*, March 1992, in *JPRS-CST*-92-013.

33. "Methods for Investigating Near-Field Power in Ground-Based High Power Laser Weapons Testing," *Zhongguo Jiguang (Chinese Journals of Lasers)*, December 1992, *JPRS-CST*-93-008.

34. One particular laser optics facility with which China hopes to cooperate is subordinated to the AGAT R&D establishment in Belarus. Chi Haotian visited AGAT in early June 1997. See "Prime Minister Linh Meets Chinese Defense Minister," *Moscow Interfax*, June 3, 1997, in *FBIS-SOV*-97-154.

35. For a comprehensive overview of the technologies associated with HPM weapons, see Carlo Kopp, "The E-Bomb—A Weapon of Electrical Mass Destruction," in Winn Schwartau, *Information Warfare*, New York: Thunder's Mouth Press, 1994, pp. 296-297. Also see J. Swegle and J. Benford, "State of the Art in High Power Microwaves: An Overview," paper presented at the Lasers 1993 International Conference on Lasers and Applications, Lake Tahoe, Nevada, December 6-10, 1993. Swegle and Benford point out that the United States, Russia, France, and the United Kingdom have HPM programs in addition to China.

36. Most technical papers on HPM are published by engineers from these three organizations. See, for example, "Numerical Simulations of Propagation of Intense Microwave Pulses," *Qiang Jiguang yu Lizishu (High Power Laser and Particle Beams)*, November 1994, in *JPRS-CST*-95-006; and "Microwave Pulse Generation," *Qiang Jiguang yu Lizishu*, May 1994, in *JPRS-CST*-94-014.

37. "Theoretical Analysis of Dual-REB Microwave Generators," *Qiang Jiguang Yu Lizi Shu*, February 1992, in *JPRS-CST*-92-015; and

Deng Yangjian, "Research on Focus Radiation Performance Microwave Directed Energy System," *Dianzi Duikang Jishu*, May 1996, pp. 1-8, in *CAMA*, 1997, Vol. 4, No. 2.

38. "Beam Energy Weaponry: Powerful as Thunder and Lightening," *Jiefangjun Bao*, December 25, 1995, in *FBIS-CHI*-96-039. Also see "Outlook for 21st Century Information Warfare," *Guoji Hangkong (International Aviation)*, March 5, 1995, in *FBIS-CHI*-95-114. The author, Chang Mengxiong, is from COSTIND's Beijing Institute of Systems Engineering and is a leading advocate for China's rapid pursuit of information warfare capabilities, to include HPM weapons. He is widely regarded as China's foremost expert of national C^4I issues. Also see Gong Jinxuan, "High Powered Microwave Weapons: A New Concept for Electronic Warfare Weaponry," *Dianzi Duikang Jishu*, February 1995, in *CAMA*, Vol. 2, No. 5. Gong is from the Southwest Institute of Electronic Equipment (SWIEE), China's premier radar ECM research institute; and Zhang Hongqi, "High Power Microwaves and Weaponry," *Xiandai Fangyu Jishu*, April 1994, pp. 38-46, in *CAMA*, Vol. 2, No. 5. Zhang is also from SWIEE.

39. Peng Guixin, "Development of Electromagnetic-Pulse Simulators," *Shijie Daodan Yu Hangtian (Missiles and Spacecraft)*, November 1990, in *JPRS-CST*-91-010. The paper was sponsored by the Beijing Institute of Electronic Engineering. Designators for the EMP simulators have been the DM-140, DM-1200, and DMF-600.

40. See Henry Freund and Robert Parker, "Free Electron Lasers," *Scientific American*, April 1989, p. 84, for the technological explanation.

41. Neil Munro, *The Quick and the Dead: Electronic Combat and Modern Warfare*, New York: St. Martin's Press, 1991, p. 50. Also see David Shukman, *Tommorrow's War: The Threat of High-Technology Weapons*, New York: Harcourt Brace, 1996, p. 207. According to Shukman, the United States has not fielded such a HPM warhead, but could within 3 years at a cost of $20 million.

42. "3-cm Relativistic Backward Wave Oscillator," *Qiang Jiguang Yu Lizishu*, May 1992, in *JPRS-CST*-92-019.

43. "More on Completion of First Domestic EMW-Pumped FEL Experimental System," *Keji Ribao*, September 9, 1992, in *JPRS-CST*-92-018.

44. "High Power Microwave Radiation From Beam-Plasma Interaction," *Qiang Jiguang Yu Lizishu*, May 1992, in *JPRS-CST*-92-016; and Wu Jianqiang, Liu Shenggang, and Mo

Yuanlong, "Plasma-Filled Dielectric Cherenkov Maser," *Qiang Jiguang yu Lizishu*, February 1997, pp. 51-58, in *FBIS-CST*-97-012.

45. Tang Changjian, Yang Zhonghai, and Liu Shenggang, *Science in China*, Vol. 36, September 1993; Yan Yang, Li Hongfu, Du Pingzhong, and Liu Shenggang, "Numerical Modeling of Large Orbit Magnetron-Type Gyrotron Through Self-Consistent Nonlinear Theory," in *IEEE Transactions on Plasma Science*, Vol 22, No. 6, December 1994, pp. 1010-1014; Li Jiayin, Xiong Xiangzheng, Liu Shenggang, "Experimental Study of Relativistic Magentron," *Qiang Jiguang yu Lizishu*, November 1997, pp. 563-567, in *FBIS-CHI*-98-128; and Ding Yaogen and Peng Jun, "Multibeam Klystron—New Type of High Power Microwave Amplifier," *Dianzi Kexue Xuekan*, January 1996, pp. 64-71, in *FBIS-CST*-96-008.

46. Yu Chuan, Li Liangzhong, Yan Chengli, Zeng Fanqun, Chen Yingshi, Cai Mingliang, Sun Chengwei, Xie Weiping, "Experimental Study of MFCG Explosive Tube," *Qiang Jiguang yu Lizishu*, November 1997, pp. 578-580, in *FBIS-CHI*-98-058. Specific SWIFP entity doing the FCG development is the Shock Wave and Detonation Physics Laboratory. Also see Dong Zhiwei, Wang Guirong, and Tian Shihong, "Performance Analysis of Helical Explosively- Driven MFCG," *Qiang Jiguang yu Lizishu*, August 1997, pp. 353-358, in *FBIS-CHI*-98-076; and Yang Libing and Gao Shunshou, "Numerical Simulation of High-Voltage Pulsed Power Conditioning Systems," *Qiang Jiguang yu Lizishu*, August 1996, pp. 400-406, in *FBIS-CHI*-97-003. Yang and Gao are from SWIFP.

47. For an overview, see Wang Ying and Ma Fuxue, *Xin Gainian Wuqi Yuanli (Principles of New Concept Weaponry)*, Beijing: Ordnance Industry Press, 1997, pp. 276-295; "Preliminary Experimental Study of Electromagnetic Missiles Conducted," *Dianzi Keji Daxue Xuebao*, February 1992, in *JPRS-CST*-92-013; "Study of Backscattering of Electromagnetic Missiles," *Dianzi Xuebao*, June 1992, in *JPRS-CST*-92-023; Ruan Chengli and Lin Weigan, "Electromagnetic Missiles and Counterstealth Technology," *Xiandai Leida*, No. 3, 1992. China's leading experts on electromagnetic missiles include Ruan Chengli, Shen Haoming, Wu Dacun (T. T. Wu), and Lin Weigan, who earned his Ph.D. from the University of California, Berkley in 1950. Wu authored one of the world's first studies on this subject in 1985; see Wu Tai Tsun, "Electromagnetic Missiles," *Journal of Applied Physics*, 1985, Vol. 57 (7), pp. 2370-2373. Also see Ruan Chengli, *Dianci Daodan Gailun (Concepts of Electromagnetic Missiles)*, Beijing: MPT Press, 1994; and Zhou Shuigeng, *Dianci Daodan Jiqi Dui Weilai Dianzizhan de Yingxiang*, ("Effect of Electromagnetic Missiles on Future Electronic Warfare") *Shanghai Hangtian*, 1994 (1), pp. 33-37.

48. "Development of Flash X-ray Machines at CAEP," *Qiang Jiguang Yu Lizishu*, August 1991, *JPRS-CST*-92-001.

49. "Flash-II: China's Accelerator Reaches World Level," *Zhejiang Ribao*, July 3, 1993, *JPRS-CST*-93-015; "Flash-II: Relativistic Electron-Beam Accelerator," *Qiang Jiguang Yu Lizishu*, August 1991, *JPRS-CST*-92-001; "Flash-II Pulsed IREB Accelerator Certified," *Renmin Ribao*, July 3, 1993, *JPRS-CST*-93-013. One of leading developers of the Flash-II at NWINT is Qiu Aici.

50. "Numerical Simulation of Vircator Oscillation for HPM Generation," *Qiang Jiguang Yu Lizishu*, August 1992, *JPRS-CST*-93-002.

51. Zhang Zhenzhou, "Directional Transmission of EMP Burst," *Xiandai Fangyu Jishu*, April 1994, pp. 47-58, in *CAMA*, Vol. 2, No. 5. Selected source of EMP burst is FCG and vircator. Zhang is from the Second Academy's Beijing Institute of Electronic Engineering.

52. Neil Munro, *The Quick and the Dead: Electronic Combat and Modern Warfare*, New York: St. Martin's Press, 1991, p. 49.

53. "Performance Improvements of 3.3 MeV LIA," *Qiang Jiguang Yu Lizishu*, May 1994, *JPRS-CST*-94-014.

54. Wang Qingsong, ed., *Xiandai Junyong Gaojishu* (*Contemporary Military Use of High Technology*), Beijing: AMS Press, 1993, pp. 139-152.

55. Ji Shifan, "Domestic and Foreign Development of Tactical Air Defense Lasers," *Zhongguo Hangtian*, December 1991. The paper was sponsored by the Beijing Institute of Electronic Engineering. Also see Wan Xuying, "Lun Zhanshu Daodan Weixie yu Duikang Cuoshi" (Discussion of the Tactical Missile Threat and Countermeasures), *Zhongguo Hangtian*, November 1992. COSTIND oversees laser and space areas of the 863 Programs. The other five 863 Program areas are managed by the State Science and Technology Commission.

56. Tong B. Tang, *Science and Technology in China*, London: Longman, 1984, p. 184. Also see Xia Zhihong, Ye Wenwei, and Cai Qingfu, "New Progress of Ranging Technology at Wuhan Satellite Laser Ranging Station," Conference Paper for the Eighth International Workshop on Laser Ranging Instrumentation at the Goddard Space Flight Center, June 1993.

57. Yang Qunfu and Liu Xiaoen, "Fanweixing Wuqi Xiting Fazhan Qianjing de Yanjiu" (Study on the Development Prospects for ASAT

Weapon Systems), *Hangtian Qingbao Yanjiu (Aerospace Information Research)*, March 1993. The U.S. MIRACL is a U.S. Navy program for ship defense against cruise missiles. During one test at White Sands, the laser achieved 2.2 MW output power with a 3800 nanometer wavelength.

58. Sun Baoju and Chen Gang, "CO_2 Radar Applications," *Zhongguo Hangtian (Aerospace China)*, February 1993.

59. The study was sponsored by China Aerospace Corporation's Sanjiang Aerospace Group, also known as 066 Base. Liu Shiliang, "Development of Foreign Counterstealth Technology," *Zhongguo Hangtian*, September 1995; on the AIOFM lidar, see "Nation's First UV Differential Absorption Lidar System Operational," *Keji Ribao*, February 18, 1997, in *FBIS-CST*-97-008.

APPENDIX V

COMMISSION OF SCIENCE, TECHNOLOGY, AND INDUSTRY FOR NATIONAL DEFENSE

The focal point for coordinating People's Liberation Army (PLA) requirements with the SSTC is the Commission of Science, Technology, and Industry for National Defense (COSTIND). COSTIND is directed by Lieutenant General Cao Gangchuan, who transfered from his previous position as director of the GSD Equipment Department. Cao has at least four deputy directors who each hold different portfolios.[1] COSTIND senior advisors include Qian Xuesen, the father of China's space and missile development (see Appendix II) and Chen Fangyun, an electronics expert who in the 1960s and 1970s organized the effort to develop a space tracking network. Important departments, committees, and institutes include the following:

- **S&T Committee.** COSTIND's S&T Committee, directed by Zhu Guangya, is made up of some of China's most talented engineers, who provide consultative guidance for the COSTIND leadership.

- **Comprehensive Planning Department.** COSTIND's Comprehensive Planning Department (*zonghe jihua bu*) is a key organization responsible for long-range planning for developing of critical enabling technologies and weapons systems.[2]

- **S&T Department.** Directed by Major General Jiang Laigen, the S&T Department likely oversees preliminary and model R&D efforts of the various COSTIND and defense industry research institutes.[3]

COSTIND also oversees the development and operation of China's space facilities and a series of engineering and test bases.[4]

- **Base 20.** Jiuquan Satellite Launch Center (JSLC), located in Gobi desert, is China's earliest and largest satellite launch center. From here, CLTC conducts medium and low earth orbit satellite launches at high inclinations. More than 20 satellites, mostly photoreconnaissance platforms, have launched from Jiuquan. Beijing has TT&C station under XSCC.

- **Base 21.** Nuclear weapons testing in Xinjiang, near the town of Lop Nur.

- **Base 22.** SAM testing near Jiuquan, Gansu province.

- **Base 23.** Naval Test area SLBM testing; antiship missiles at Huludao, Liaoning province. Established in 1957 for joint Sino-Soviet anti-ship missile testing.

- **Base 25.** Taiyuan Space Launch Center, in northern Shanxi province in the area of Kelan, Wuzhai, and Xingxian counties, one of China's primary satellite launch and missile testing sites. Commanded by Major General Jiang Xuefu.

- **Base 26.** Xian Satellite Control Center is directed by Major General Li Hengxing, who recently replaced Major General Shangguan Shipan.

- **Base 27.** Xichang Space Launch Center, directed by Major General Hu Shixiang.

- **Base 29.** China Aerodynamic Research and Development Center (CARDC) in Mianyang, Sichuan province.

- **Base 31.** Baicheng Conventional Weapons Testing Center.

China Launch and Tracking Control General (CLTC). CLTC (*Weixing Fashe Cekong Xitong Zongbu*) is a departmental level entity which functions as an operational arm of COSTIND. CLTC, commanded by Major General Li Yuanzheng, is roughly analagous to U.S. Space Command. With over 5000 engineers and technicians, CLTC oversees three satellite launch centers and China's vast tracking, telemetry, and control (TT&C) network. CLTC headquarters, which occupies a large facility on Beijing's North Third Ring Road, is made up of a planning department, development department, tracking and control department, logistics department and liaison department. Two research institutes under CLTC include the Luoyang Institute of Tracking, Telecommunications, Technology (LITTT) and the Beijing Special Engineering and Design Research Institute (BSEDI).[5] The COSTIND entity responsible for constructing launch and tracking sites and testing facilities is the Beijing Special Engineering and Design Research Institute (BSEDI).[6]

ENDNOTES - APPENDIX V

1. Deputy Directors include Lieutenant General Zhang Xuedong, Lieutenant General Shen Rongjun, Lieutenant General Huai Guomo, Lieutenant General Shen Chunnian, Major General Chen Dazhi, and Major General Wang Tongye. Portfolios include conventional systems R&D and testing; space and missile systems R&D (*Shen Rongjun*); telecommunications (*Wang Tongye*); nuclear issues; and defense conversion (*Huai Guomo*). At the time of this writing, much of COSTIND was being placed under a new general department, the General Armaments Department, co-equal to the GSD, GPD, and GLD.

2. At least since 1991, the Comprehensive Planning Department has been directed by Major General Chen Dazhi. Chen, however, has been recently promoted to become a deputy director of COSTIND. See Xinhua press release, March 31, 1997, in *FBIS-CHI*-97-091.

3. Organizational information drawn from "Directory of PRC Military Personalities," Defense Liaison Office, U.S. Consulate General, Hong Kong, October 1996, pp. 25-27. For an excellent overview of COSTIND, see Shirley A. Kan, *China: Commission of Science Technology, and Industry for National Defense*, Congressional Research Service Report for Congress, November 7, 1996.

4. Lewis and Xue, pp. 85, 189, 191, and 279 endnote 64; and "Directory of PRC Military Personalities," pp. 25-27. Bases within China's military industrial complex can be divided into series. For example, the 20 and 30 series are COSTIND bases for space launch and

weapons testing. The 50 series are Second Artillery strategic missile bases. The 60 series are subordinate to China Aerospace Corporation, China's space and missile industry.

5. Undated CLTC brochure; "Directory of PRC Military Personalties," Defense Liaison Office, U.S. Consulate General, Hong Kong, October 1996, pp. 25-27.

6. Undated BSEDI brochure. Founded in 1958, BSEDI has over 450 personnel.

APPENDIX VI

CHINA'S MINISTRY OF ELECTRONICS INDUSTRY (MEI)

Minister: Hu Qili

Vice-ministers: Liu Jianfeng (dual-hatted as Chairman of Liantong. Next in line to replace Hu Qili)
Zhang Jinqiang
Lu Xinkui

Chief Engineer: Yu Zhongyu

China Academy of Electronics and Information Technology (CAEIT). General systems design department. Heavily involved in formation of national information infrastructure. Works closely with COSTIND's Beijing Institute of Systems Engineering (BISE) in developing China's national C^4I infrastructure.

Research Institutes.

2nd Research Institute	Taiyuan
5th Research Institute	Associated with environmental testing.
6th Research Institute	Computer systems engineering. Also known as Huasun Computer Company.
7th Research Institute	Guangzhou Communications Research Institute. Has conducted R&D on field mobile communications systems, such as digital mobile communications system.
8th Research Institute	Anhui Fiber Optical Fiber Research Institute.

10th Research Institute	Southwest Institute of Electronics Technology (SWIET). Located in Chengdu. Involved in a range of defense-related programs including UHF, microwave, and millimeter communications and radar equipment.
11th Research Institute	R&D into solid state laser systems, to include laser range finders.
12th Research Institute	TACAN systems.
13th Research Institute	Located in Shijiazhuang. Involved in development of integrated circuits and solid state lasers. Best known for work in gallium arsenide integrated circuits. Importing advanced French technology.
14th Research Institute	Leading radar institute, located in Nanjing. Involved in development of early warning, phased array, HF, and space tracking radars.
15th Research Institute	North China Computer Institute. Located in Beijing and known as Taiji.
18th Research Institute	Tianjin Institute of Power Sources.
20th Research Institute	China's primary navigation research institute located in Xian.
21st Research Institute	Located in Shanghai.
22nd Research Institute	China Institute of Radiowave Propagation. Involved in timing

	sources, such as those associated with Shaanxi Astronautical Observatory Timing Station.
25th Research Institute	R&D on signal processing systems. Working on long wave infrared (LWIR) imaging seeker.
26th Research Institute	Located in Chongqing. R&D into surface acoustic wave (SAW) devices, piezoelectronic acousto-optics, electronic ceramics, and crystals.
28th Research Institute	Nanjing Research Institute of Electrical Engineering. Responsible for C^4I systems integration for air defense and air traffic control systems.
29th Research Institute	Southwest Institute of Electronic Engineering (SWIEE). Located in Chengdu. Responsible for radar reconnaissance and electronic countermeasures.
30th Research Institute	R&D on switching systems, to include advanced common channel signalling seven (SS7) software.
33rd Research Institute	Located in Taiyuan.
34th Research Institute	Guilin Institute of Optical Communications. One of China's principle entities engaged in R&D on fiber optics. JV with Nokia.
36th Research Institute	Responsible for communications electronic countermeasures.
38th Research Institute	East China Institute of Electronic Engineering

	(ECRIEE). Specializes in early warning and artillery radars. Located in Hefei.
39th Research Institute	Northwest Institute of Electronic Equipment (NWIEE). Develops SATCOM ground stations, microwave relays, and missile range equipment.
40th Research Institute	Responsible for connectors and relays. Located in Bengbu.
41st Research Institute	R&D into signal generators and test equipment for infrared focal plane arrays.
43rd Research Institute	Hengli Electronics Development Corporation. Located in Hefei.
44th Research Institute	Chongqing Institute of Optoelectronics. R&D into charged couple devices (CCDs), infrared focal plane arrays, and fiber optics.
45th Research Institute	R&D into integrated circuit production technology, i.e., steppers. Located in Pingliang, Gansu province.
46th Research Institute	Located in Tianjin. R&D and testing of silicon and gallium arsenide materials.
47th Research Institute	R&D into reduced instruction set computing (RISC) integrating circuits.
49th Research Institute	Northeast Institute of Sensor Technology. Located in Harbin. Develops vibration and other sensors.

50th Research Institute	Shanghai Institute of Microwave Technology. Has worked on automated command systems for SAM units.
51st Research Institute	Conducts R&D on radar reconnaissance and jamming.
53rd Research Institute	Institute of Applied Infrared Technology. Located in Liaoning. Engaged in passive jamming and optoelectronics.
54th Research Institute	Communications Technology Institute. Located in Shijiazhuang. Major player in wide range of military systems.
55th Research Institute	R&D on semiconductors.

Factories.

605th Factory	Fiber optic cable.
701 Factory	Radios
707 Factory	Chenxing Radio Factory
710 Factory	Zhongyuan Radio Factory. Located in Wuhan. Engaged in production of civilian and military telecommunications equipment.
711 Factory	Shipborne UHF systems.
712 Factory	Airborne UHF systems. Located in Tianjin.
713 Factory	unknown
714 Factory	Panda Electronics Factory. HF and airborne UHF systems.
716 Factory	Digital communications equipment

719 Factory	Airborne navigation equipment
720 Factory	One China's primary radar factories. Works closely with 14th Research Institute in Nanjing.
722 Factory	ECM plant. Works closely with 29th Research Institute.
730 Factory	Submarine cable
734 Factory	Fiber optic cable and wireless equipment
738 Factory	Computers. Works closely with 15th Research Institute.
741 Factory	Optoelectronics and infrared systems.
750 Factory	Guangdong Radio Group Telecommuications Company. HF SSB.
754 Factory	Located in Tianjin.
756 Factory	Aviation navigation equipment
760 Fatory	Troposcatter systems
761 Factory	Beijing Broadcast Factory. High powered VLF systems.
764 Factory	Tianjin Broadcasting Equipment Company. Aviation navigation equipment
765 Factory	Aviation navigation equipment. Located in Baoji.
769 Factory	Airborne UHF systems
780 Factory	Airborne radar countermeasures
781 Factory	ECM plant and bombing control radars

782 Factory	Airborne radars and transponders. Located in Baoji.
783 Factory	Fujiang Machinery Factory. Also known as Sichuan Jiuzhou Electronic Factory. Produces secondary radars and IFF transponders. Located in Mianyang.
784 Factory	Jinjiang Electronic Machinery Factory. Located in Chengdu. Produces surveillance radars.
785 Factory	Produces optoelectronics equipment and SAM guidance radars, and AAA computers.
786 Factory	SAM guidance radars. Located in Xian.
789 Factory	AAA computers
834 Factory	Tactical communications equipment
913 Factory	ECM plant. Works closely with 36th Research Institute.
914 Factory	Lanxin Radio Factory. Located in Lanzhou.
924 Factory	Radar reconnaissance and jamming equipment. Works closely with 29th Research Institute.
4500 Factory	Computer systems
4508 Factory	Located in Tianjin.
6909 Factory	ECM plant

Other Entities

Nanjing Radio Factory	Tactical communications systems.
Nanhai Machinery Factory	Airborne navigation radars
Huajing Electronics Group	Working on submicron integrated circuits under 908 project. Located in Wuxi. Joint venture with AT&T.
Huadong Computer Technology Institute	Developed China's first minisupercomputer.
Great Wall Computer Group	Largest computer firm in China.
Beijing Institute of Electronic Technology Applications	Developed Internet firewall system.
North China Research Institute of Opto-Electronics	Works on sub-systems for lidar system. Develops infrared materials and devices.
Zhongchen Electronics Industry Development Corporation	Engaged in systems engineering. JV with US Lotus Corporation.
Zhongruan Corporation	Computer software.

China's Top Integrated Circuit Producers:

Shougang-NEC

Tianjin-Motorola

Huajing Microelectronics Group

Shanghai Beijing Microelectronics Company

Shanghai Advanced Semiconductor Corporation

Shaoxing Huayue Microelectronics

Sources: Various *Foreign Broadcast Information Service (FBIS)* reports, 1995-present; *China Today: Defense Science and Technology*, Vol. II, pp. 719-764; author's unpublished study, *Ministry of Electronics Research Institutes and Factories: Organizations Engaged in Research and Development of Command, Control, Communications, and Intelligence (C^3I)*, U.S. Defense Attache Office, Beijing, May 1995; and Lynn Crisanti, *Report on Ministry of Electronics Industry Research Institutes and Factories*, Kamsky Associates, 1998.

ABOUT THE AUTHOR

MARK A. STOKES is Country Director for the PRC and Taiwan within the Office of the Secretary of Defense, International Security Affairs (OSD/ISA). Major Stokes has previous assignments as a signals intelligence and electronic combat support officer in the Philippines and West Berlin. He served as assistant air attaché at the U.S. Defense Attaché Office in Beijing, China, from 1992-95. Before assignment to OSD/ISA, he was the Asia-Pacific regional planner within the HQ USAF Operations and Plans Directorate from 1995-97. He holds graduate degrees in International Relations and Asian Studies from Boston University and the Naval Postgraduate School. He received his formal Chinese Mandarin language training from the Defense Language Institute in Monterey, California, and the Diplomatic Language Services in Rosslyn, Virginia.